# Soldier's Manual of Common Tasks and Warrior Skills Level 2, 3, and 4

## Department of Defense

Pentagon
Publishing

*STP 21-24-SMCT

Soldier Training Publication
No. 21-24-SMCT

Headquarters
Department of the Army
Washington, DC, 9 September 2008

# SOLDIER'S MANUAL OF COMMON TASKS WARRIOR LEADER SKILLS LEVEL 2, 3, and 4

## TABLE OF CONTENTS



**Distribution Restriction:** Approved for public release; distribution is unlimited.

*This manual supersedes STP 21-24-SMCT, 2 October 2006.

**Skill Level 3**

# PREFACE

This manual is one of a series of Soldier training publications that support individual training. Commanders, trainers, and Soldiers will use this manual and STP 2-1-SMCT, *Soldier's Manual of Common Tasks, Warrior Skills Level 1*, to plan, conduct, sustain, and evaluate individual training of warrior tasks and battle drills in units.

This manual contains an Army Warrior Training plan for warrior leader skills level (SL) 2 through SL 4 and task summaries for SL 2 critical common tasks that support unit wartime missions. This manual is the only authorized source for these common tasks. Task summaries in this manual supersede any common tasks appearing in MOS-specific Soldier manuals.

Training support information such as reference materials, websites, ammunition requirements, and reproducible evaluation forms are also included. Trainers and first-line supervisors will ensure that SL 2 through SL 4 Soldiers have access to this publication in their work areas, unit learning centers, and unit libraries.

This manual applies to the Active Army, the Army National Guard (ARNG)/Army National Guard of the United States (ARNGUS), and the U.S. Army Reserve (USAR) unless otherwise stated.

The proponent of this publication is the United States Army Training and Doctrine Command (TRADOC), with the United States Army Training Support Center (ATSC) designated as the principle publishing, printing, and distribution agency. Proponents for the specific tasks are the Army schools and agencies as identified by the school code, listed in appendix A. This code consists of the first three digits of the task identification number.

Record any comments or questions regarding the task summaries contained in this manual on a DA Form 2028 (*Recommended Changes to Publications and Blank Forms*) and send it to the respective task proponent with information copies forwarded to—

- Commander, U.S. Army Training and Doctrine Command, ATTN: ATTG-I, Fort Monroe, VA 23651-5000.
- Commander, U.S. Army Training Support Center, ATTN: ATIC-ITSC-CM, Fort Eustis, VA 23604-5166.

Chapter 1

# Introduction to the SMCT System

## 1-1. GENERAL

This manual contains the critical common tasks for warrior skills levels 2, 3, and 4. Mastering the performance of these tasks will help the individual Soldier and the Soldiers that he or she supervises fight better and survive on the battlefield, and perform across the full spectrum of operations. Each Soldier must be able to perform all common critical tasks for that skill level and below. The individual Soldier shares responsibility with the trainer to sustain the skills and knowledge required to perform all warrior skills and warrior leader skills.

**a.** ***Individual Soldiers.*** To ensure that you can perform each task, have another Soldier or your supervisor periodically evaluate your performance using the task summaries. If you have questions about how to perform a task or which task to perform, ask your first-line supervisor. The first-line supervisor knows how to perform each task and can direct you to the appropriate training materials. You should also check the Reimer Digital Library for new training materials. It is your responsibility to use these materials to maintain your proficiency.

**b.** ***Trainers.*** If you are a supervisor and trainer and have Soldiers working for you, you must train them to do the tasks for their skill level and below. Commanders and trainers should use the Soldier's Manual of Common Tasks (SMCT), military occupational specialty (MOS) specific Soldier's training publications (STPs), and mission training plans (MTPs) to establish effective training plans and programs which integrate individual and collective tasks.

## 1-2. SMCT AND SUPPORT OF BATTLE-FOCUSED TRAINING

**a.** ***Overview.*** The SMCTs document the common tasks by skill level on which all Soldiers must be trained and evaluated. These documents assist leaders in identifying the strengths and weaknesses of the Soldiers in their unit and provide a means for evaluating the effectiveness of the unit's individual training program. Evaluation results identify Soldiers who need training on specific tasks. Evaluation results also tell leaders where to concentrate training to improve unit readiness and help Soldiers develop professionally.

**b.** ***Leader's Assessment.*** The leader's assessment focuses on specific unit mission requirements. It gives commanders a way to evaluate Soldier performance on individual tasks that directly support their unit mission. These tasks may be common tasks as well as MOS-specific tasks. Leader's assessments should be conducted year round. Unit commanders should make leader's comments an integral part of their unit training so that hands-on evaluation is systematically performed at the unit level.

**(1)** Tasks selected for leader's assessments include, but are not limited to, individual tasks that—

**(a)** Support the units mission essential task list (METL).

**(b)** Support other non-METL unit tasks as shown in the MTP.

**(c)** Are identified by higher headquarters for inclusion in planned individual training.

**(d)** Were rated as substandard on previous training feedback, such as the common task test (CTT) or annual general inspections.

**(e)** Are relevant to the Soldier's MOS but not required in his or her current duty assignment.

**(2)** Leaders may conduct their assessment—

**(a)** Before, during, or after individual skills training.

**(b)** As part of MTP training.

**(c)** On the job.

**(d)** At specially prepared test sites or in a battlefield scenario.

**(e)** During training or job breaks.

**(f)** After hours in the barracks.

**(g)** During special squad or individual competitive events.

**(3)** The leader's assessment evaluates the combat effectiveness of Soldiers and the unit. Commanders can use this evaluation to correct training deficiencies and plan unit training. Commanders can also use the results for personnel actions (such as preparing enlisted evaluation reports, making recommendations for promotions, and other personnel management decisions).

**c.** ***Army Warrior Training (AWT).*** AWT is hands-on training focused on the warrior tasks and battle drills learned in initial entry training (IET). AWT is based on individual and leader tasks described in STP 21-1-SMCT and STP 21-24-SMCT. Units may conduct AWT *any time* during the collective training period and should be integrated with training exercises to conserve resources and to improve realism. For example, units could test Soldiers during—

**(1)** MTP evaluations and other collective training activities.

**(2)** Weapons qualification.

**(3)** Gas chamber training.

**(4)** Mission rehearsal exercises at Combat Training Centers (CTCs).

**(5)** Reception, staging, onward movement, and integration (RSOI) training in theater.

***Note***: AWT results are objective measures for the commander to use to evaluate unit readiness and the effectiveness of the training program. Leaders also consider AWT results when preparing enlisted evaluation reports and recommending Soldiers for promotion.

**d.** ***Evaluation of Performance Measures.*** The GO/NO GO performance measures for SL2 through SL4 task summaries are included in this manual. Supervisors can use the performance measures at the end of each task summary to evaluate Soldier skill proficiencies.

## 1-3. SMCT FEATURES

**a.** ***AWT Training Plan.*** The Army warrior training plan (chapter 2) lists, by skill level, the critical common tasks for which all Soldiers are responsible. It indicates where each task is first taught to standard and how often training on the task is required to sustain proficiency. Leaders should use this information to develop a comprehensive unit training plan.

**b.** ***Task Summary.*** Each task summary documents the performance requirements of a warrior task. The summaries provide the Soldier and the trainer with the information necessary to evaluate the critical tasks. The task summaries use the following format:

**(1) Task Title**. The task title identifies the action to be performed.

**(2) Task Number**. A 10-digit number identifies each task. The first three digits of the number represent the proponent code for that task. (A list of the proponent codes is given in appendix A.) Include the entire 10-digit task number, along with the task title, in any correspondence relating to the task.

**(3) Conditions.** The task conditions identify all the equipment, tools, materials, references, job aids, and supporting personnel that the Soldier needs to perform the task. This section identifies environmental conditions that can alter task performance (such as visibility, temperature, or wind). This section also identifies specific cues or events (such as a chemical attack or identification of an unexploded ordnance hazard) that trigger task performance.

**(4) Standards.** A task standard specifies the requirements for task performance by indicating how well, completely, or accurately a product must be produced, a process must be performed, or both. Standards are described in terms of accuracy, tolerances, completeness, format, clarity, number of errors, quantity, sequence, or speed of performance.

**(5) Training and Evaluation Guide.** This section has two parts. The first part, Performance Steps, lists the individual steps that the Soldier must complete to perform the task. The second part is the Performance Evaluation Guide. This provides guidance on how to evaluate a Soldier's performance of the task. It is composed of three subsections. The Evaluation Preparation subsection identifies special setup procedures and, if required, instructions for evaluating the task performance. Sometimes the conditions and standard must be modified so that the task can be evaluated in a situation that does not exactly duplicate actual field performance. This subsection may also include instructions that the evaluator should give to the Soldier before the performance test. The Performance Measures subsection identifies the criteria for acceptable task performance. The Soldier is rated (GO/NO GO) on each specific action or specific product produced. The Soldier must score a GO on all (or each specified) performance measure to receive a GO on the task.

**(6) References.** This section identifies references that provide more detailed and thorough explanations of task performance requirements than that given in the task summary description. This section identifies resources that the Soldier can use to improve or maintain performance.

Additionally, task summaries can include safety statements, environmental considerations, and notes. Safety statements (danger, warning, caution) alert users to the possibility of immediate death, personal injury, or damage to equipment. Notes provide additional information to support task performance.

**c.** ***Training Support.*** This manual includes the following that provide additional training support information.

**(1)** Appendix A (Proponent School and Agency Codes) lists the task proponents and agency codes (first three digits of the task number) with addresses for submitting comments concerning specific tasks in this manual.

**(2)** Appendix B (Guide to Forms) explains the use of various SMCT training and evaluation forms and, in the online version, provides links to the forms.

(3) Glossary lists abbreviations and acronyms and their definitions.

(4) References lists all reference materials cited in the task summaries by type, identification number, title, and date.

## 1-4.CONDUCTING ARMY WARRIOR TRAINING AND EVALUATION

a. ***Role of the Commander.*** As a commander, you must ensure that your training plan prepares the unit for the full spectrum of operations. The plan should enable your Soldiers to develop and sustain proficiency on the MOS-specific and warrior tasks for their skill level. Use critical common task summaries to evaluate your Soldiers' proficiency on those tasks critical to your unit mission. An effective training program converts unproductive time into effective training time. This will upgrade the skills of individual Soldiers and promote the development of junior leaders. To develop an effective unit training program we reiterate, from the STP 21-1-SMCT, the following seven-step approach—

***Step 1.*** Set the objectives for training.

***Step 2.*** Plan the resources (personnel, time, funds, facilities, devices, training aids).

***Step 3.*** Train the trainers.

***Step 4.*** Provide the resources.

***Step 5.*** Manage risks, environmental and safety considerations.

***Step 6.*** Conduct the training.

***Step 7.*** Evaluate the results.

b. ***Role of the Trainer.*** Although training is everyone's business, NCOs are the key to training the individual Soldiers assigned to the unit. NCOs should be the first to recognize which tasks each Soldier can or cannot perform. NCOs must ensure that each Soldier takes steps to master these tasks. This manual will assist the NCO (the trainer) in doing what trainers do best—train. To train effectively, NCOs must perform the following functions.

**(1) Plan the training.** You can usually integrate or conduct training for specific common tasks concurrently with other training or during slack periods. Use the Army Warrior Training Plan in chapter 2 to identify the critical warrior tasks for which each Soldier is responsible.

**(2) Prepare yourself.** Get training guidance from your chain of command on when to train, what to train, and what resources are available. Know the training objectives for each task and ensure that you can perform the task. Gather the necessary training references for each task as listed in the task summary.

**(3) Obtain the resources.** Gather the required resources and prepare the training site according to the conditions statement and the evaluation preparation section of the task summary. Ensure that equipment needed to complete the task is operational. Coordinate the use of training aids and devices.

**(4) Train the Soldiers.** Show the Soldiers how to do the task to standard and explain each step. Give each Soldier at least one chance to perform the task.

**(5) Evaluate the Soldiers.** Evaluate how well the Soldiers perform. You may conduct the evaluation during individual training or while evaluating individual performance while conducting unit collective tasks. Use the Army Warrior Training Plan to determine how often to check Soldiers to ensure they maintain proficiency.

(6) **Record the results.**

**(a)** Score the Soldier GO if all performance measures are passed. Score the Soldier NO GO if any step is failed. If the Soldier fails any step, show or tell him or her what was done wrong and how to do it correctly.

**(b)** Record the GO/NO GO results in the leader book. You may use DA Form 5165-R; see the Guide to Forms (appendix B) at the end of this publication. Do not make written entries directly on the evaluation guides in the SMCT.

**(c)** When applicable, conduct an after-action review exercise to allow training participants to discover for themselves what happened, why it happened, and how it can be done better. Once all key points have been discussed and linked to future training, the evaluator will make the appropriate notes for inclusion into the score.

**(7) Retrain and reevaluate.** Work with Soldiers until they can perform the task to specific Soldier's manual standard.

---

*Good training increases the professionalism of each Soldier and helps to develop an efficient unit. You are a vital link in the conduct of good training.*

---

Chapter 2

# Training Guide

## 2-1. THE ARMY WARRIOR TRAINING PLAN

The AWT plan provides information to help the trainer plan, prepare, train, evaluate, and monitor individual training in units. It lists by general subject area, skill level and warrior tasks or battle drill that the critical common task Soldiers must perform, the initial training location, and a suggested frequency of training. The training location column uses brevity codes to indicate where the task is first taught to standards. If the task is taught in the unit the word, "UNIT" appears in this column. Tasks trained via self-development media are indicated by "SD." If it is taught in the training base, the brevity code (BCT, OSUT, and AIT) of the resident course appears. Brevity codes and resident courses are listed below.

| Brevity Codes | |
|---|---|
| ANCOC | Advanced NCO Course |
| BNCOC | Basic NCO Course |
| WLC | Warrior Leaders Course |
| BCT | Basic Combat Training |
| OSUT | One Station Unit Training |
| AIT | Advanced Individual Training |
| UNIT | Trained in / by the Unit |

The sustainment training column lists how often (frequency) Soldiers should train on the task to ensure they maintain their proficiency. This information is a guide for commanders to develop a comprehensive unit-training plan. The commander, in conjunction with the unit trainers, is in the best position to determine on which tasks and how often Soldiers need training to maintain unit readiness.

| Frequency Codes | |
|---|---|
| AN | Annually |
| SA | Semiannually |
| QT | Quarterly |

| Army Warrior Training Plan | | | |
|---|---|---|---|
| ***Task Number*** | ***Title*** | ***Training Location*** | ***Sustainment Training Frequency*** |
| **Warrior Leader Skill Level 2** | | | |
| **Subject Area 1. Individual Conduct and Laws of War** | | | |
| **158-100-3006** | Resolve an Ethical Problem | Unit | AN |
| **181-101-2023** | Enforce the Uniform Code of Military Justice (UCMJ) | Unit | AN |
| **181-105-2001** | Enforce the Law of War and the Geneva and Hague Conventions | Unit | AN |
| **181-105-2002** | Conduct Combat Operations According to the Law of War | Unit | AN |
| **224-176-2426** | Enforce Compliance with Media Ground Rules | Unit | AN |
| **805C-PAD-2401** | The Army's EO Program and Sexual Harassment Policy | Unit | SA |
| **Subject Area 2. First Aid** | | | |
| **081-831-0101** | Request Medical Evacuation | WLC | QT |
| **081-831-1056** | Coordinate Medical Activity Support | Unit | SA |
| **081-831-1057** | Supervise Compliance with Preventive Medicine Measures (PMM) | Unit | AN |
| **Subject Area 3. Chemical, Biological, Radiological, and Nuclear (CBRN)** | | | |
| **031-503-1001** | Identify Chemical Agents Using M256-Series Chemical Agent Detector Kit | Unit | AN |
| **031-503-1002** | Conduct Unmasking Procedures | Unit | AN |
| **031-503-1005** | Submit a Chemical, Biological, Radiological and Nuclear (CRBN) Report | WLC | AN |

| Army Warrior Training Plan | | | |
|---|---|---|---|
| *Task Number* | *Title* | *Training Location* | *Sustainment Training Frequency* |
| **031-503-1010** | Supervise the Employment of Chemical, Biological, Radiological and Nuclear Markers | Unit | AN |
| **031-503-2053** | Report Chemical, Biological, Radiological and Nuclear (CBRN) Information Using NBC 4 | WLC | AN |
| **031-504-1061** | Conduct a Mask Fit Test Using the M41 Protection Assessment Test System (PATS) | Unit | AN |
| **Subject Area 4. Survive (Combat Techniques)** | | | |
| **061-283-6003** | Adjust Indirect Fire | WLC | AN |
| **071-326-5705** | Establish an Observation Post | WLC | AN |
| **071-410-0019** | Control Organic Fires | Unit | AN |
| **Subject Area 5. Navigate** | | | |
| **071-329-1019** | Use a Map Overlay | Unit | AN |
| **071-326-0515** | Select a Movement Route Using a Map | WLC | SA |
| **Subject Area 6: Communicate** | | | |
| **158-100-4003** | Communicate Effectively at the Direct Leadership Level | Unit | AN |
| **158-100-4009** | Communicate in Writing | Unit | AN |
| **Subject Area 22. Unit Operations** | | | |
| **071-326-5502** | Issue a Fragmentary Order | WLC | QT |
| **071-326-5503** | Issue a Warning Order | WLC | QT |
| **071-730-0006** | Enforce Operations Security | Unit | AN |
| **Subject Area 23. Security and Control** | | | |
| **301-371-1200** | Process Captured Materiel | Unit | AN |

| Army Warrior Training Plan | | | |
|---|---|---|---|
| ***Task Number*** | ***Title*** | ***Training Location*** | ***Sustainment Training Frequency*** |
| **Subject Area 25. Equipment Checks** | | | |
| **091-CTT-2001** | Supervise Preventive Maintenance Checks and Services (PMCS) | Unit | AN |
| **Subject Area 27. Risk Managment** | | | |
| **153-001-2000** | Employ CRM Process and Principles and Show How They Apply to Performance of My Job/Assigned Duties | WLC | AN |
| **Subject Area 28. Administration/Management** | | | |
| **805C-PAD-2060** | Report Casualties | WLC | AN |
| **805C-PAD-2044** | Recommend Individual for Award | Unit | AN |
| **805C-PAD-2145** | Counsel a Soldier on Noncommissioned Officer Evaluation Report Contents and Checklist | Unit | AN |
| **805C-PAD-2407** | Recommend Disciplinary Action for a Soldier | Unit | AN |
| **805C-PAD-2503** | Enforce Compliance with the Army's Equal Opportunity and Sexual Harassment Policies | WLC | SA |
| **Subject Area 29: Build Teamwork** | | | |
| **158-100-3012** | Correlate a Leader's Role in Character Development with Values and Professional Obligation | Unit | AN |
| **158-100-7003** | Counsel a Subordinate | Unit | AN |
| **158-100-7012** | Develop Subordinates | Unit | AN |
| **158-100-7015** | Develop an Effective Team | Unit | AN |
| **158-100-8006** | Solve Problems Using Military Decisionmaking Process | Unit | AN |

| Army Warrior Training Plan | | | |
|---|---|---|---|
| ***Task Number*** | ***Title*** | ***Training Location*** | ***Sustainment Training Frequency*** |
| **Warrior Leader Skill Level 3** | | | |
| **Subject Area 2. First Aid** | | | |
| **081-831-1058** | Supervise Casualty Treatment and Evacuation | Unit | AN |
| **081-831-1059** | Implement Measures to Reduce Combat Stress | Unit | AN |
| **Subject Area 3. Chemical, Biological, Radiological, and Nuclear (CBRN)** | | | |
| **031-503-1016** | Implement Mission-Oriented Protective Posture (MOPP) | Unit | AN |
| **031-503-3004** | Supervise the Crossing of a Contaminated Area | Unit | AN |
| **Subject Area 4. Survive (Combat Techniques)** | | | |
| **052-195-3066** | Direct Construction of Nonexplosive Obstacles | BNCOC | AN |
| **071-331-0820** | Analyze Terrain | Unit | AN |
| **071-410-0012** | Conduct Occupation of an Assembly Area | BNCOC | |
| **071-420-0021** | Conduct a Movement to Contact by a Platoon | Unit | AN |
| **Subject Area 5. Navigate** | | | |
| **071-332-5000** | Prepare an Operational Overlay | BNCOC | AN |
| **Subject Area 20. Defense Measures** | | | |
| **071-430-0002** | Conduct a Defense by a Squad | Unit | AN |
| **Subject Area 22. Unit Operation** | | | |
| **071-326-5805** | Conduct a Route Reconnaissance Mission | Unit | AN |
| **071-332-5021** | Prepare a Situation Map | BNCOC | QT |
| **551-88M-3601** | Perform duties as Convoy Commander | Unit | AN |
| **551-88N-3042** | Plan Unit Move | Unit | SA |
| **805C-PAD-3594** | Store Classified Information and Materials | Unit | AN |

| Army Warrior Training Plan | | | |
|---|---|---|---|
| *Task Number* | *Title* | *Training Location* | *Sustainment Training Frequency* |
| **Subject Area 23. Security and Control** | | | |
| **301-371-1052** | Protect Classified Information and Material | Unit | AN |
| **191-379-4407** | Plan Convoy Security Operations | Unit | AN |
| **Subject Area 24. Enemy Personnel** | | | |
| **191-377-4250** | Supervise the Processing of Detainees at the Point of Capture | Unit | AN |
| **191-377-4252** | Supervise the Escort of Detainees | Unit | AN |
| **191-377-4203** | Supervise the Establishment and Operation of a Roadblock/Checkpoint | Unit | AN |
| **191-378-5315** | Supervise an Installation Access Control Point | Unit | AN |
| **191-378-4303** | Supervise a Riot/Crowd Control Operation with Squad Size Element | Unit | AN |
| **Subject Area 27. Risk Management** | | | |
| **153-001-3001** | Control Mission Safety Hazard | BNCOC | QT |
| **Warrior Leader Skill Level 4** | | | |
| **Subject Area 3: Chemical, Biological, Radiological and Nuclear (CBRN)** | | | |
| **031-503-4002** | Supervise Unit Preparation for a Chemical, Biological, Radiological and Nuclear Attack | Unit | AN |
| **Subject Area 4. Survive (Combat Techniques)** | | | |
| **071-326-5775** | Coordinate with an Adjacent Platoon | Unit | QT |
| **Subject Area 20. Defense Measures** | | | |
| **071-430-0006** | Conduct a Defense by a Platoon | ANCOC | QT |

| Army Warrior Training Plan | | | |
|---|---|---|---|
| *Task Number* | *Title* | *Training Location* | *Sustainment Training Frequency* |
| **Subject Area 22. Unit Operations** | | | |
| **071-326-3013** | Conduct a Tactical Road March | Unit | AN |
| **071-720-0015** | Conduct an Area Reconnaissance by a Platoon | ANCOC | AN |
| **091-CTT-4001** | Supervise Maintenance Operations | Unit | AN |
| **Subject Area 23. Security and Control** | | | |
| **191-379-4440** | Supervise the Evacuation of Dislocated Civilians | Unit | SA |
| **191-379-5403** | Supervise a Riot/Crowd Control Operation with a Platoon-Sized Element | Unit | SA |
| **191-379-4408** | Plan Security for a Command Post | Unit | SA |
| **Subject Area 26. Crime Prevention** | | | |
| **191-379-4425** | Implement the Unit's Crime Prevention Program | Unit | AN |
| **Subject Area 27. Risk Management** | | | |
| **153-001-4001** | Integrate Risk Management into Mission Plans | Unit | AN |
| **Subject Area 28. Administration/Management** | | | |
| **805C-PAD-4359** | Manage Soldier's Deployment Requirements | Unit | SA |
| **805C-PAD-4597** | Integrate Newly Assigned Soldiers | Unit | QT |
| **805C-PAD-4550** | Prepare a Standing Operating Procedure (SOP) | Unit | SA |
| **805C-PAD-2472** | Prepare a Duty Roster | Unit | SA |

## 2-2. SUBJECT AREA CODES

Below is a list of subject areas contained in this STP.

| Warrior Leader Skills Level 2, 3, and 4 | |
|---|---|
| 1 | Individual Conduct and Laws of War |
| 2 | First Aid |
| 3 | Chemical, Biological, Radiological, and Nuclear |
| 4 | Survive (Combat Techniques) |
| 5 | Navigate |
| 6 | Communicate |
| 20 | Defense Measures |
| 22 | Unit Operations |
| 23 | Security and Control |
| 24 | Enemy Personnel |
| 25 | Equipment Checks |
| 26 | Crime Prevention |
| 27 | Risk Management |
| 28 | Administration/Management |

## Skill Level 2

## SUBJECT AREA 1: INDIVIDUAL CONDUCT AND LAWS OF WAR

# 158-100-3006

**Resolve an Ethical Problem**

**Conditions:** Given FM 5-0, FM 6-22, DOD 5500.7-R, and a situation which requires you to make an ethical decision.

**Standards:** Implement a solution based upon sound reasoning and judgment in the application of an abbreviated military problem solving process to an ethical problem.

| Performane Steps |
|---|
| **1.** Define the ethical problem. |
| **2.** Identify applicable laws, regulations and values. |
| **3.** Develop acceptable possible solutions. |
| **4.** Implement the best ethical solution. |

**Evaluation Preparation:** *Setup:* Provide the Soldier with the references listed below. Prepare a scenario that requires the Soldier to respond accurately, according to task standards, to the following performance measures. This may be presented orally or in writing.

*Brief Soldier:* Tell the Soldier that he or she will be required to correctly respond to all of the performance measures to receive a GO on the task.

| Performance Measures | GO | NO GO |
|---|---|---|
| **1.** Defined the ethical problem. | ___ | ___ |
| **2.** Identified all applicable laws, regulations, and values. | ___ | ___ |
| **a.** Indentified applicable laws and rules. | | |
| **b.** Correctly identified all appropriate ethical values. | | |
| **c.** Accurately determined all relevant guiding moral principles from the ethical values chosen. | | |
| **3.** Developed acceptable possible solutions that addressed the problem. | ___ | ___ |
| **a.** Developed possible solutions. | | |
| **b.** Accurately applied all relevant information to the ethical problem. | | |

| Performance Measures | GO | NO GO |
|---|---|---|
| **4.** Implemented the best ethical solution. | ___ | ___ |
| **a.** Chose a solution which reflected sound judgment and analysis. Selected the best method to implement the solution. | | |
| **b.** Selected the best method to implement the solution. | | |
| **c.** Accurately assessed results and modified the plan as appropriate. | | |

**Evaluation Guidance:** Refer to chapter 1, paragraph 1-4 b (6).

**References**

**Required:** FM 6-22

**Related:** FM 1-0, FM 5-0, FM 6-0, FM 6-22 (FM 22-100), FM 7-21.13, and TC 1-05

## 181-101-2023

### Enforce the Standards of the Uniform Code of Military 004Austice (UCMJ)

**Conditions:** You are a noncommissioned officer (NCO) in a leadership position in the U.S. Army. You are responsible for understanding that disciplinary action against a Soldier for misconduct is a command responsibility. You are responsible for understanding the military justice system, including the Uniform Code of Military Justice (UCMJ) and disciplinary options available to a commander. You are responsible for identifying potential violations of the UCMJ and expeditiously reporting them to the appropriate authorities for investigation and processing. You have access to AR 600-20, AR 27-10, and FM 27-10.

**Standards**: Understand that disciplinary action against a Soldier for misconduct is a command responsibility. Understand the military justice system, including the UCMJ and the disciplinary options available to a commander. Identify potential violations of the UCMJ and expeditiously report them to theappropriate authorities for investigation and processing.

| Performance Steps |
|---|

**1.** Describe how disciplinary action against a Soldier is a command responsibility.

**2.** Identify who has authority to take disciplinary action against a Soldier for misconduct.

**3.** Describe a Soldier's responsibility to identify potential or actual violations of the UCMJ and expeditiously report these violations to the appropriate authorities for investigation and processing.

**4.** Describe a commander's responsibility to conduct a preliminary investigation into misconduct allegedly committed by a Soldier under his/her command.

**Performance Steps**

**a.** Describe the basis and procedures of a commander's inquiry.

**b.** Describe the basis and procedures of an AR 15-6 investigation.

**c.** Describe the requirement for the military police or Criminal Investigation Command (CID) to conduct a criminal investigation.

**5.** List the disciplinary options available to the commander.

**a.** Describe how a commander can take no action at all or close a case.

**b.** Describe how a commander can use administrative or nonpunitive measures.

**(1)** List administrative or nonpunitive disciplinary measures available to a commander.

**(2)** Describe why a commander would use nonpunitive or administrative disciplinary measures rather than impose nonjudicial punishment or proceed to court-martial.

**(3)** Describe how a NCO leader may be involved in imposing nonpunitive or administrative disciplinary measures, such as counseling or corrective training, to a subordinate Soldier.

**c.** Describe how a commander can use nonjudicial punishment.

**(1)** Define nonjudicial punishment.

**(2)** List the different types of nonjudicial punishment.

**(3)** Describe nonjudicial punishment procedures.

**(4)** Describe a Soldier's legal rights during nonjudicial punishment procedures.

**(a)** Formal proceedings.

**(b)** Summarized proceedings.

**(5)** List the maximum punishment available under nonjudicial punishment.

**(a)** Formal proceedings.

**(b)** Summarized proceedings.

**(6)** Describe a Soldier's appellate rights under nonjudicial punishment.

**(a)** Formal proceedings.

**(b)** Summarized proceedings.

**d.** Describe how a commander can use judicial punishment.

**(1)** Define judicial punishment.

**(2)** List the different types of court-martial in the military justice system.

**(3)** Describe judicial or court-martial procedures.

**(4)** Describe a Soldier's legal rights during judicial or court-martial punishment.

**Performance Steps**

**(5)** List the maximum punishment available under judicial or court-martial punishment.

**(6)** Describe a Soldier's appellate rights under judicial or court-martial punishment.

**6.** List factors a commander should consider when determining what disciplinary option to pursue.

**a.** Describe whether a commander should consider the character and military service of the accused.

**b.** Describe whether a commander should consider the nature and circumstances of the offense and the extent of the harm caused.

**c.** Describe whether a commander should consider the needs of the Service and the probable effect of his/her decision on the command and the military community.

**d.** Describe whether a commander should consider the disposition of similar offenses in the past and the general disciplinary trends within the command.

**e.** Describe whether a commander should consider the appropriateness of the authorized punishment to the particular accused and offense.

**f.** Describe whether a commander should determine whether he/she has jurisdiction over the accused and the offense.

**g.** Describe whether a commander should consider the availability and admissibility of evidence against the accused.

**h.** Describe whether a commander should consider the cooperation of the accused in the apprehension or conviction of others.

**i.** Describe whether a commander should consider the possible improper motives of the accuser.

**j.** Describe whether a commander should consider that the victim or others are reluctant to testify.

**7.** Describe the permissibility of a commander discussing and gaining input from an NCO leader regarding which disciplinary option to pursue against a subordinate Soldier within the unit.

**8.** Describe the authority of an NCO to issue a lawful order to a subordinate Soldier.

**a.** Describe the duty of a subordinate Soldier to follow this order.

**b.** Describe the potential adverse ramifications for a Soldier violating this order.

**c.** Describe the elements and maximum punishment available under Article 91, UCMJ.

**Evaluation Preparation:** *Setup:* Evaluate this task at the end of military justice training.

*Brief Soldier:* Tell the Soldier that he/she will be evaluated on his/her ability to understand that disciplinary action against a Soldier for misconduct is a command responsibility. Tell the Soldier that he/she will also be evaluated on his/her ability to understand the military justice system, including the UCMJ; the disciplinary options available to a commander; and the ability to identify potential violations of the UCMJ, and expeditiously report them to the appropriate authorities for investigation and processing.

| Performance Measures | GO | NO GO |
|---|---|---|
| **1.** Described how disciplinary action against a Soldier is a command responsibility. | —— | —— |
| **2.** Identified who had authority to take disciplinary action against a Soldier for misconduct. | —— | —— |
| **3.** Described a Soldier's responsibility to identify potential or actual violations of the UCMJ and expeditiously report these violations to the appropriate authorities for investigation and processing. | —— | —— |
| **4.** Described a commander's responsibility to conduct a preliminary investigation into misconduct allegedly committed by a Soldier under his/her command. | —— | —— |
| **a.** Described the basis and procedures of a commander's inquiry. | | |
| **b.** Described the basis and procedures of an AR 15-6 investigation. | | |
| **c.** Described the requirement for the military police or CID to conduct a criminal investigation. | | |
| **5.** Listed disciplinary options available to the commander. | —— | —— |
| **a.** Described how a commander could take no action at all or close a case. | | |
| **b.** Described how a commander could use administrative or nonpunitive measures. | | |
| **(1)** Listed administrative or nonpunitive disciplinary measures available to a commander. | | |
| **(2)** Described why a commander would use nonpunitive or administrative disciplinary measures rather than impose nonjudicial punishment or proceed to court-martial. | | |
| **(3)** Described how a NCO leader may be involved in imposing nonpunitive or administrative disciplinary measures, such as counseling or corrective training, to a subordinate Soldier. | | |

| Performance Measures | GO | NO GO |
|---|---|---|
| **c.** Described how a commander could use nonjudicial punishment.<br>**(1)** Defined nonjudicial punishment.<br>**(2)** Listed the different types of nonjudicial punishment.<br>**(3)** Described nonjudicial punishment procedures.<br>**(4)** Described a Soldier's legal rights during nonjudicial punishment procedures.<br>**(5)** Listed the maximum punishment available under nonjudicial punishment.<br>**(6)** Described a Soldier's appellate rights under nonjudicial punishment.<br>**d.** Described how a commander could use judicial punishment.<br>**(1)** Defined judicial punishment.<br>**(2)** Listed the different types of court-martial in the military justice system.<br>**(3)** Described judicial or court-martial procedures.<br>**(4)** Described a Soldier's legal rights during judicial or court-martial punishment.<br>**(5)** Listed the maximum punishment available under judicial or court-martial punishment.<br>**(6)** Described a Soldier's appellate rights under judicial or court-martial punishment. | | |
| **6.** Listed factors a commander should consider when determining what disciplinary option to pursue.<br>**a.** Described whether a commander should consider the character and military service of the accused.<br>**b.** Described whether a commander should consider the nature and circumstances of the offense and the extent of the harm caused.<br>**c.** Described whether a commander should consider the needs of the Service and the probable effect of his/her decision on the command and the military community.<br>**d.** Described whether a commander should consider the disposition of similar offenses in the past and the general disciplinary trends within the command.<br>**e.** Described whether a commander should consider the appropriateness of the authorized punishment to the particular accused and offense. | ___ | ___ |

| Performance Measures | GO | NO GO |
|---|---|---|
| **f.** Described whether a commander should determine whether he/she has jurisdiction over the accused and the offense. | | |
| **g.** Described whether a commander should consider the availability and admissibility of evidence against the accused. | | |
| **h.** Described whether a commander should consider the cooperation of the accused in the apprehension or conviction of others. | | |
| **i.** Described whether a commander should consider the possible improper motives of the accuser. | | |
| **j.** Described whether a commander should consider that the victim or others are reluctant to testify. | | |
| **7.** Described the permissibility of a commander discussing and gaining input from an NCO leader regarding which disciplinary option to pursue against a subordinate Soldier within the unit. | ___ | ___ |
| **8.** Described the authority of an NCO to issue a lawful order to a subordinate Soldier. | ___ | ___ |
| **a.** Described the duty of a subordinate Soldier to follow this order. | | |
| **b.** Described the potential adverse ramifications for a Soldier violating this order. | | |
| **c.** Described the elements and maximum punishments available under Article 91, UCMJ. | | |

**Evaluation Guidance:** Refer to chapter 1, paragraph 1-4 b (6).

**References**

**Required:** AR 27-10 and AR 600-20

**Related:**

# 181-105-2001

## Enforce the Law of War and the Geneva and Hague Conventions

**Conditions:** You are a Soldier in the U.S. Army. As a Soldier, you are responsible for identifying, understanding, and complying with the provisions of the Law of War, including the Geneva and Hague Conventions. You are also responsible for identifying and notifying the appropriate authorities of any suspected or known violations of the Law of War. The appropriate authorities, including your chain of command, must enforce the provisions of the Law of War, including the Geneva and Hague Conventions.

**Standards:** Identify, understand, and comply with the Law of War. Identify problems or situations that violate the policies and take appropriate action, including notifying appropriate authorities, so that expedient action may be taken to correct the problem or situation.

| Performance Steps |
|---|

**1.** Identify the key violations of the Law of War.

**a.** Define a Law of War violation.

**b.** Identify the two general types of war crimes.

**c.** Define what constitutes a grave breach and give examples.

**d.** Describe what obligations exist if there is a grave breach.

**e.** Define what constitutes an other than grave breach or a simple breach and give examples.

**f.** Describe what obligations exist if there is a simple breach.

**2.** Describe the responsibilities of U.S. Soldiers to obey the Law of War.

**a.** Describe how U.S. Soldiers are bound to obey all the rules of the Customary Law of War and the Hague and Geneva Conventions.

**b.** Describe how U.S. Soldiers may be court-martialed for violating these rules.

**c.** Describe how U.S. Soldiers may also be prosecuted for committing a war crime.

**3.** Describe the responsibilities of the commander in regards to violations of the Law of War.

**a.** Describe how the legal responsibility for the commission of a war crime can be placed on the commander as well as the subordinate who actually commits the war crime.

**b.** Describe the circumstances under which a commander may be prosecuted for the commission of a war crime.

**4.** Describe a criminal order and a Soldier's responsibility toward a criminal order.

**a.** Describe the applicability of a Soldier asserting the defense of "obeying a superior order" for the commission of a war crime.

**b.** Describe whether a subordinate Soldier, who actually commits a war crime, is excused from prosecution if the commander is charged with the commission of the war crime.

**c.** Describe the responsibility of a Soldier to disobey any order which requires the Soldier to commit criminal acts in violation of the Law of War.

| Performance Steps |
|---|
| **d.** Describe the responsibility of a Soldier to obey the rules of engagement (ROE) and the potential consequences for violating the ROE. |
| **e.** Describe the responsibility of the Soldier to ask a superior for clarification of an order presumed to be criminal or illegal. |
| **5.** Identify the key requirements in processing violations of the Law of War. |
| **a.** Describe the actions required when a Law of War violation is suspected. |
| **b.** Describe your combatant commander-in-chief's (CINC) guidance. |
| **c.** Describe the requirement to report suspected violations of the Law of War. |
| **d.** Describe the obligations of the investigative team. |
| **(1)** Describe who appoints the investigating officer(s). |
| **(2)** Describe the qualifications of the investigating officer(s). |
| **(3)** Describe the importance of timely collection of information/evidence. |
| **(4)** Describe the initial report format. |

**Evaluation Preparation**: *Setup*: Evaluate this task at the end of Law of War training.

*Brief Soldier*: Tell the Soldier that he/she will be evaluated on his/her ability to identify, understand, and comply with the Law of War, including the Geneva and Hague Conventions. Tell the Soldier that he/she will also be evaluated on his/her ability to identify problems or situations that violate the Law of War and take appropriate action, including notifying appropriate authorities of actual or suspected violations, so that expedient action may be taken to correct the problem or situation.

| Performance Measures | GO | NO GO |
|---|---|---|
| **1.** Identified the key violations of the Law of War. | —— | —— |
| **a.** Defined a Law of War violation. | | |
| **b.** Identified the two general types of war crimes. | | |
| **c.** Defined what constituted a grave breach and gave examples. | | |
| **d.** Described what obligations existed if there was a grave breach. | | |
| **e.** Defined what constituted an other than grave breach or a simple breach and gave examples. | | |
| **f.** Described what obligations existed if there was a simple breach. | | |

| Performance Measures | GO | NO GO |
|---|---|---|
| **2.** Described the responsibilities of U.S. Soldiers to obey the Law of War. | ___ | ___ |
| **a.** Described how U.S. Soldiers were bound to obey all the rules of the Customary Law of War and the Hague and Geneva Conventions. | | |
| **b.** Described how U.S. Soldiers may be court-martialed for violating these rules. | | |
| **c.** Described how U.S. Soldiers may also be prosecuted for committing a war crime. | | |
| **3.** Described the responsibilities of the commander in regards to violations of the Law of War. | ___ | ___ |
| **a.** Described how the legal responsibility for the commission of a war crime could be placed on the commander as well as the subordinate who actually committed the war crime. | | |
| **b.** Described the circumstances under which a commander may be prosecuted for the commission of a war crime. | | |
| **4.** Described a criminal order and a Soldier's responsibility toward a criminal order. | ___ | ___ |
| **a.** Described the applicability of a Soldier asserting the defense of "obeying a superior order" for the commission of a war crime. | | |
| **b.** Described whether a subordinate Soldier, who actually committed a war crime, was excused from prosecution if the commander was charged with the commission of the war crime. | | |
| **c.** Described the responsibility of a Soldier to disobey any order which required the Soldier to commit criminal acts in violation of the Law of War. | | |
| **d.** Described the responsibility of a Soldier to obey the ROE and the potential consequences for violating the ROE. | | |
| **e.** Described the responsibility of the Soldier to ask a superior for clarification of an order presumed to be criminal or illegal. | | |

| Performance Measures | GO | NO GO |
|---|---|---|
| **5.** Identified the key requirements in processing violations of the Law of War. | ___ | ___ |
| **a.** Described what actions were required when a Law of War violation was suspected. | | |
| **b.** Described your combatant CINC's guidance. | | |
| **c.** Described the requirement to report suspected violations of the Law of War. | | |
| **d.** Described the obligations of the investigative team. | | |
| **(1)** Described who appointed the investigating officer(s). | | |
| **(2)** Described the qualifications for the investigating officer(s). | | |
| **(3)** Described the importance of timely collection of information/evidence. | | |
| **(4)** Described the initial report format. | | |

**Evaluation Guidance:** Refer to chapter 1, paragraph 1-4 b (6).

**References**

**Required:** AR 27-1, FM 27-10, FM 27-2, and TC 27-10-1

**Related:**

## 181-105-2002

### Conduct Combat Operations According to the Law of War

**Conditions:** You are a Soldier in a deployed unit which has a mission that requires you and your subordinates to be actively involved in operations that are governed by the Law of War.

**Standards:** Conduct your operations according to the Law of War and employ actions to prevent violations of the Law of War.

| Performance Steps |
|---|

**1.** Identify the key elements of the Hague and Geneva Conventions that pertain to combat operations.

**a.** Unlawful and lawful targets.

**b.** Noncombatants.

**c.** Lawful use of force.

**d.** Protected property—property dedicated to the humanities, structures, and items of cultural or historical significance, and schools, orphanages and other places dedicated to the use and benefit of children.

**e.** Protected medical transports and facilities.

**Performance Steps**

**f.** Proper use of medical service symbols, a flag of truce, national emblems, and/or insignia or uniforms of an opposing force.

**g.** Define perfidy and treachery.

**h.** Proportionate use of force.

**i.** List weapons and action that causes unnecessary suffering and harm.

**j.** List duties of the captor to prisoners of war (POWs), retained persons, and detainees.

**k.** List duties of combatants toward civilians and civilian property.

**l.** List rights, responsibilities, and discipline of POWs.

**m.** List war crimes constituting grave breaches.

**n.** List State obligations when evidence of a war crime exists.

**o.** List requirements to report Law of War violations.

**2.** Employ actions to prevent violations of the Law of War.

**a.** Protect the following:

**(1)** Noncombatants/civilians.
**(2)** Property.
**(3)** POWs, retained persons, and detainees.
**(4)** Medical transports and facilities.

**b.** Prevent engagement of unlawful targets.

**(1)** Protective emblems.
**(2)** Noncombatants and protected property.

***Note*:** Do not fire indiscriminately.

**(3)** Rules of engagement (ROE).

**c.** List actions to prevent excessive use of force.

**d.** List actions to prevent unauthorized use of medical service symbols, flags of truce, national emblems, and enemy uniform/insignia.

**e.** List actions to prevent unnecessary destruction or seizure of property.

**f.** List actions to prevent unnecessary suffering and harm.

**g.** List actions to enforce the rights and responsibilities of POWs.

**h.** Describe the obligations of a military commander with respect to prevention, suppression, and prosecution of war crimes.

| Performance Steps |
|---|
| **i.** Describe the concept of command responsibility in relation to war crimes. |
| **j.** List methods for reporting violations of the Law of War. |

**Evaluation Preparation:** *Setup:* Evaluate this task at the end of Law of War training.

*Brief Soldier:* Tell the Soldier that he/she will be evaluated on his/her ability to conduct combat operations according to the Law of War and to employ actions to prevent violations of the Law of War.

| Performance Measures | GO | NO GO |
|---|---|---|
| **1.** Employed actions that prevented Law of War violations and war crimes to protect— | ___ | ___ |
| **a.** Noncombatants and civilians. | | |
| **b.** Property. | | |
| **c.** POWs and detainees. | | |
| **d.** Medical transports and facilities. | | |
| **2.** Employed actions that prevented Law of War violations and war crimes: | ___ | ___ |
| **a.** Engagement of lawful targets. | | |
| **b.** Excessive use of force. | | |
| **c.** Unauthorized use of medical services symbol, flag of truce, national emblems, and enemy insignias and uniforms. | | |
| **d.** Unnecessary suffering and harm. | | |
| **3.** Employed actions that enforce the rights and responsibilities of POWs. | ___ | ___ |
| **4.** Listed the duties of the captor to POWs, retained persons, and detainees. | ___ | ___ |
| **5.** Listed the duties of combatants toward civilians and civilian property. | ___ | ___ |
| **6.** Listed the rights and responsibilities of POWs. | ___ | ___ |
| **7.** Listed war crimes that constitute grave breaches. | ___ | ___ |
| **8.** Listed the State obligations that occurred when evidence of a war crime existed. | ___ | ___ |
| **9.** Listed the requirement to report Law of War violations. | ___ | ___ |

| Performance Measures | GO | NO GO |
|---|---|---|
| **10.** Listed the specific ROE. | —— | —— |
| **11.** Listed the commander's primary responsibilities. | —— | —— |

**Evaluation Guidance:** Refer to chapter 1, paragraph 1-4 b (6).

**References**

**Required:** AR 27-1, FM 27-10, FM 27-2, and TC 27-10-1

**Related:**

## 224-176-2426

### Enforce Compliance with Media Ground Rules

**Conditions:** Given a copy of your command/installation's media ground rules in a field, military operations in urban terrain (MOUT) or garrison environment, enforce compliance with the rules.

**Standards:** Identify the steps for enforcing media ground rules.

**CAUTION**

Media ground rules may change between various command levels within garrison, field and deployment environments.

**Performance Steps**

*Note*: Ground rules recognize the importance of media coverage of military operations and are not intended to prevent release of derogatory, embarrassing, or negative information. However, during operations, specific information on friendly troop movements, tactical deployments, and dispositions could jeopardize operational security and endanger lives.

*Note*: Commanders and public affairs practitioners will establish basic ground rules ensuring the free flow of information while safeguarding classified materials or operational plans.

1. Obtain area-specific ground rules from the local public affairs office or representative.

2. Define media ground rules. Media representatives are required to contact public affairs prior to conducting any media-related activities.

3. Determine if the media representative is aware of the area ground rules; if not, explain the violations.

4. Be professional.

5. Inform the chain of command and public affairs representative if ground rules are violated.

**Evaluation Preparation:** *Setup*: Provide Soldier with a copy of the media ground rules. Obtain ground rules from the local public affairs office (PAO), chain of command or refer to FM 46-1. Have one Soldier play the part of a media representative.

*Brief Soldier*: Tell the Soldier media representatives are in the area and want to interview Soldiers for a print or broadcast story. Evaluate the Soldier on his/her ability to enforce media ground rules.

| Performance Measures | GO | NO GO |
|---|---|---|
| 1. Ensured that the Soldier knew where to obtain a copy of the media ground rules. | ___ | ___ |
| 2. Ensured that the Soldier could define media ground rules. | ___ | ___ |
| 3. Briefed media representative of violations to the established ground rules. | ___ | ___ |
| 4. Reported ground rule violations to chain of command or public affairs representative. | ___ | ___ |

**Evaluation Guidance:** Refer to chapter 1, paragraph 1-4 b (6).

**References**

**Required:** AR 360-1

**Related:** FM 46-1

## 805C-PAD-2401

### The Army's Equal Opportunity Program and Sexual Harassment Policy

**Conditions:** You are a leader in a unit or section responsible for supervising personnel. The personnel include both male and female and represent different races, colors, religions and national origins. You have access to AR 600-20, TC 26-6, AR 670-1, AR 690-12, AR 690-600, and DOD 1350.2. The following agencies or individuals are available for assistance: equal opportunity advisor (EOA), equal opportunity leader (EOL), chaplain, staff judge advocate (SJA), provost marshal (PM), inspector general (IG), and Chief, Community Housing Referral and Relocation Services (CHRRS) Office.

**Standards:** Ensure that unit leaders communicate the Army's equal opportunity (EO) program and the sexual harassment policy. Take appropriate actions to prevent discrimination based on the five protected EO categories and prevent sexual harassment in units. Take immediate and appropriate actions to resolve sexual harassment omplaints. Take immediate and appropriate actions to resolve EO complaints. Use support agencies and/or personnel to facilitate implementing all aspects of the Army's EO policies. Keep the chain of command informed of all actions that involve complying with the Army's EO policy.

***Note***: All above actions must be completed in the above prescribed manner.

| Performance Steps |
|---|

**1.** Ensure that Soldiers in your organization are familiar with the Army's EO policy:

**a.** The U.S. Army will provide equal opportunity and fair treatment for military personnel and family members without regard to race, color, religion, gender, or national origin, and provide an environment free from unlawful discrimination and offensive behavior. This policy applies—

**(1)** On and off post.

**(2)** To working, living, and recreational environments (including both on and off post housing).

**Performance Steps**

**b.** Soldiers will not be accessed, classified, trained, assigned, promoted, or otherwise managed on the basis of race, color, religion, gender, or national origin.

***Note***: The assignment and utilization of female Soldiers is governed by federal law. AR 600-13, prescribes policies, procedures, responsibilities and the position coding system for female Soldiers.

**c.** Requires rating and reviewing officials to evaluate each member's commitment to the elimination of unlawful discrimination and/or sexual harassment and to document significant deviations from that commitment in evaluation reports.

**d.** Substantiated formal complainants require a "Does not support EO" on the NCOER or the OER. This documentation includes administering appropriate administrative, disciplinary, or legal actions(s) to correct inappropriate behavior.

**e.** Physical disability and age controls are not addressed due to overriding concerns of medical fitness and deplorability of military personnel.

**2.** Ensure that Soldiers know the five protected categories (cornerstones) within the Army's EO policy as listed in step 1 above (race, color, religion, gender, or national origin).

**3.** Inform Soldiers of the Army's EO program related elements.

Related EO elements are subject areas that can, and often do, include actions that are usually based on discrimination or prejudicial behaviors. While not directly linked to EO, when individuals take inappropriate action in these areas, it often results in discrimination that is based on the cornerstones of the EO program, race, color, religion gender or national origin.

- Relationships between Soldiers of different rank.
- Consideration of others.
- Army language policy.
- Accommodating religious practices.
- Tattoos.
- Extremist organizations.
- Army values.

**4.** Identify the reasons for conducting an EO climate assessment.

**a.** Regulatory: All company, troop, and battery equivalent commanders are required to conduct a climate and training needs assessment within 90 days of assuming command (180 days for reserve component) and annually thereafter.

**b.** Command Directed: The commander, at any level, in attempting to evaluate the execution of his or her EO program may direct a climate assessment.

| Performance Steps |
|---|

**c.** Driven by Events: This type of assessment is normally conducted to ascertain the cause and effect relationship precipitated by major EO events, trends, or other unfavorable conditions within an organization.

**d.** Staff Assistance Visit: During a staff assistance visit, the EO staff advisor or other staff personnel may conduct a climate assessment to provide the commander with feedback status and execution of the command's EO program.

**e.** Monitor EO Action Plan: Climate assessment can also be used to monitor the impact or success of the unit's EO Action Plan (EOAP) goals or milestones.

**5.** Ensure that Soldiers are aware of the definition of sexual harassment.

> Sexual harassment is a form of gender discrimination that involves unwelcomed sexual advances, requests for sexual favors, and other verbal or physical conduct of a sexual nature between the same or opposite genders when—
>
> - Submission to, or rejection of, such conduct is made either explicitly or implicitly a term or condition of a person's job, pay, career; or
> - Submission to, or rejection of, such conduct by a person is used as a basis for career or employment decisions affecting that person; or
> - Such conduct has the purpose or effect of unreasonable interfering with an individual's work performance or creates an intimidating, hostile or offensive.

**6.** Identify the types of sexual harassment.

**a.** Quid Pro Quo: A Latin term meaning "this for that." This term refers to conditions placed on a person's career or term of employment in return for sexual favors.

**b.** Hostile Environment: This occurs when Soldiers or civilians are subjected to offensive, unwanted, and unsolicited comments and behaviors of a sexual nature. If these behaviors have the potential of unreasonably interfering with their performance, then the environment is classified as hostile.

**Evaluation Preparation:** *Setup:* Provide the Soldier with the references listed in the conditions statement. Prepare a scenario that requires the Soldier to respond accurately according to task standards to the following performance measures.

*Brief Soldier:* Tell the Soldier that he/she will have to correctly perform all of the performance measures in order to receive a GO.

| Performance Measures | GO | NO GO |
|---|---|---|
| **1.** Accurately discussed the Army's EO policy. | ____ | ____ |
| **2.** Correctly identified the five protected EO categories. | ____ | ____ |

| Performance Measures | GO | NO GO |
|---|---|---|
| **3.** Correctly identified five of the seven Army's EO program related elements. | ___ | ___ |
| **4.** Correctly identified three of the five reasons for conducting a climate assessment. | ___ | ___ |
| **5.** Correctly defined the Army's sexual harassment policy. | ___ | ___ |
| **6.** Correctly identified the two types of sexual harassment. | ___ | ___ |

**Evaluation Guidance:** Refer to chapter 1, paragraph 1-4 b (6).

**References**

**Required:** AR 600-20, TC 26-6, AR 670-1, AR 690-12, AR 690-600, and DOD 1350.2.

**Related:**

**SUBJECT AREA 2: FIRST AID**

## 081-831-0101

### Request Medical Evacuation

**Conditions:** You have a casualty requiring medical evacuation (MEDEVAC) and you will need a patient pickup site. You will need operational communications equipment, MEDEVAC request format, a standard scale military map, a grid coordinate scale, and unit signal operation instructions (SOIs).

**Standards:** Transmit a MEDEVAC request, providing all necessary information within 25 seconds. Transmit, as a minimum, line numbers 1 through 5 during the initial contact with the evacuation unit. Transmit lines 6 through 9 while the aircraft or vehicle is en route, if not included during the initial contact.

**Performance Steps**

1. Collect all applicable information needed for the MEDEVAC request.

   **a.** Determine the grid coordinates for the pickup site. (See STP 21-1-SMCT, task 071-329-1002.)

   **b.** Obtain radio frequency, call sign, and suffix.

   **c.** Obtain the number of patients and precedence.

   **d.** Determine the type of special equipment required.

   **e.** Determine the number and type (litter or ambulatory) of patients.

   **f.** Determine the security of the pickup site.

   **g.** Determine how the pickup site will be marked.

   **h.** Determine patient nationality and status.

**Performance Steps**

**i.** Obtain pickup site nuclear, biological, and chemical (NBC) contamination information, normally obtained from the senior person or medic.

***Note:*** NBC line 9 information is only included when contamination exists.

**2.** Record the gathered MEDEVAC information using the authorized brevity codes. (See table 081-831-0101-1.)

***Note***: Unless the MEDEVAC information is transmitted over secure communication systems, it must be encrypted, except as noted in step 3b(1).

**a.** Location of the pickup site (line 1).

**b.** Radio frequency, call sign, and suffix (line 2).

**c.** Numbers of patients by precedence (line 3).

**d.** Special equipment required (line 4).

**e.** Number of patients by type (line 5).

**f.** Security of the pickup site (line 6).

**g.** Method of marking the pickup site (line 7).

**h.** Patient nationality and status (line 8).

**i.** NBC contamination (line 9).

**3.** Transmit the MEDEVAC request. (See STP 21-1-SMCT, task 113-571-1022.)

**a.** Contact the unit that controls the evacuation assets.

**(1)** Make proper contact with the intended receiver.

**(2)** Use effective call sign and frequency assignments from the SOI.

**(3)** Give the following in the clear "I HAVE A MEDEVAC REQUEST;" wait one to three seconds for a response. If no response, repeat the statement.

**b.** Transmit the MEDEVAC information in the proper sequence.

**(1)** State all line item numbers in clear text. The call sign and suffix (if needed) in line 2 may be transmitted in the clear.

***Note***: Line numbers 1 through 5 must always be transmitted during the initial contact with the evacuation unit. Lines 6 through 9 may be transmitted while the aircraft or vehicle is en route.

**(2)** Follow the procedure provided in the explanation column of the MEDEVAC request format to transmit other required information.

**(3)** Pronounce letters and numbers according to appropriate radio/telephone procedures.

**(4)** Take no longer than 25 seconds to transmit.

**(5)** End the transmission by stating "Over."

**(6)** Keep the radio on and listen for additional instructions or contact from the evacuation unit.

## Performance Steps

### Table 081-831-0101-1. MEDEVAC request format

| LINE | ITEM | EXPLANATION | WHERE/HOW OBTAINED | WHO NORMALLY PROVIDES | REASON |
|---|---|---|---|---|---|
| 1 | Location of Pickup Site. | Encrypt the grid coordinates of the pickup site. When using the DRYAD Numeral Cipher, the same "SET" line will be used to encrypt the grid zone letters and the coordinates. To preclude misunderstanding, a statement is made that grid zone letters are included in the message (unless unit SOP specifies its use at all times). | From Map | Unit Leader(s) | Required so evacuation vehicle knows where to pick up patient. Also, so that the unit coordinating the evacuation mission can plan the route for the evacuation vehicle (if the evacuation vehicle must pick up from more than one location). |
| 2 | Radio Frequency, Call Sign, and Suffix | Encrypt the frequency of the radio at the pickup site, not a relay frequency. The call sign (and suffix if used) of the person to be contacted at the pickup site may be transmitted in the clear. | From SOI | RTO | Required so that the evacuation vehicle can contact the requesting unit while enroute (obtain additional information or change in situation or directions). |
| 3 | Number of Patients by Precedence | Report only applicable information and encrypt the brevity codes.<br>A - URGENT.<br>B - URGENT-SURG.<br>C - PRIORITY.<br>D - ROUTINE.<br>E - CONVENIENCE.<br>If two or more categories must be reported in the same request, insert the word "BREAK" between each category. | From Evaluation of Patient(s) | Medic or Senior Person Present | Required by the unit controlling the evacuation vehicles to assist in prioritizing missions. |
| 4 | Special Equipment Required | Encrypt the applicable brevity codes.<br>A - None.<br>B - Hoist.<br>C - Extraction equipment.<br>D - Ventilator. | From Evaluation of the Patient/ Situation | Medic or Senior Person Present | Required so that the equipment can be placed on board the evacuation vehicle prior to the start of the mission. |
| 5 | Number of Patients by Type | Report only applicable information and encrypt the brevity code. If requesting MEDEVAC for both types, insert the word "BREAK" between the litter entry and ambulatory entry.<br>L + # of Pnt - Litter<br>A + # of Pnt - Ambulatory (sitting) | From Evaluation of Patient(s) | Medic or Senior Person Present | Required so that the appropriate number of evacuation vehicles may be dispatched to the pickup site. They should be configured to carry the patients requiring evacuation. |
| 6 | Security of the Pickup Site (Wartime) | N - No enemy troops in the area.<br>P - Possibly enemy troops in the area (approach with caution).<br>E - Enemy troops in the area (approach with caution).<br>X - Enemy troops in the area (armed escort required). | From Evaluation of the Situation | Unit Leader | Required to assist the evacuation crew in assessing the situation and determining if assistance is required. More definitive guidance can be furnished to the evacuation vehicle while it is en route (specific location of the enemy to assist an aircraft in planning its approach). |

**Performance Steps**

**Table 081-831-0101-1. MEDEVAC request format (continued)**

| LINE | ITEM | EXPLANATION | WHERE/HOW OBTAINED | WHO NORMALLY PROVIDES | REASON |
|---|---|---|---|---|---|
| 6 | Number and Type of Wound, Injury, or Illness (Peacetime) | Specific information regarding patient wounds by type (gunshot or shrapnel). Report serious bleeding, along with patient blood type, if known. | From Evaluation of Patient | Medic or Senior Person Present | Required to assist evacuation personnel in determining treatment and special equipment needed. |
| 7 | Method of Marking Pickup Site | Encrypt the brevity codes.<br>A - Panels.<br>B - Pyrotechnic signal.<br>C - Smoke signal.<br>D - None.<br>E - Other. | Based on the Situation and Availability of Materials | Medic or Senior Person Present | Required to assist the evacuation crew in identifying the specific location of the pickup. Note that the color of the panels or smoke should not be transmitted until the evacuation vehicle contacts the unit (just prior to its arrival). For security, the crew should identify the color and the unit verify it. |
| 8 | Patient Nationality and Status | The number of patients in each category need not be transmitted. Encrypt only the applicable brevity codes.<br>A - US military.<br>B - US civilian.<br>C - Non-US military.<br>D - Non-US civilian.<br>E - EPW | From Evaluation of Patient | Medic or Senior Person Present | Required to assist in planning for destination facilities and need for guards. Unit requesting support should ensure that there is an English-speaking representative at the pickup site. |
| 9 | CBRN Contamination (Wartime) | Include this line only when applicable. Encrypt the applicable brevity codes.<br>C - Chemical.<br>B - Biological.<br>R - Radiological.<br>N - Nuclear. | From the Situation | Medic or Senior Person Present | Required to assist in planning for the mission. (Determine which evacuation vehicle will accomplish the mission and when it will be accomplished.) |
| 9 | Terrain Description (Peacetime) | Include details of terrain features in and around the proposed landing site. If possible, describe relationship of the site to prominent terrain feature (lake, mountain, tower). | From an Area Survey | Personnel at Site | Required to allow evacuation personnel to assess route/avenue of approach into the area. Of particular importance if hoist operation is required. |

**Evaluation Preparation:** *Setup:* Evaluate this task during a training exercise involving a MEDEVAC aircraft or vehicle, or simulate it by creating a scenario and providing the information as the Soldier requests it. You or an assistant will act as the radio contact at the evacuation unit during "transmission" of the request. Give a copy of the MEDEVAC request format to the Soldier.

*Brief Soldier*: Tell the Soldier to prepare and transmit a MEDEVAC request. State that the communication net is secure.

| Performance Measures | GO | NO GO |
|---|---|---|
| **1.** Collected all information needed for the MEDEVAC request line items 1 through 9. | —— | —— |
| **2.** Recorded the information using the authorized brevity codes. | —— | —— |
| **3.** Transmitted the MEDEVAC request within 25 seconds. | —— | —— |

**Evaluation Guidance:** Refer to chapter 1, paragraph 1-4 b (6).

**References**

**Required:**

**Related:** FM 8-10-6

## 081-831-1056

### Coordinate Medical Acitivity Support

**Conditions:** You are a noncommissioned officer (NCO) deployed in a forward unit. You are directed to coordinate support from available health care services.

**Standards:** Integrate combat health support functional areas into the military decisionmaking process (MDMP). Coordinate the force health protection plan with the battalion medical platoon. Ensure that a field sanitation team program is established. Coordinate the preventive medicine (PVTMED) program through the medical platoon to reduce incidents of disease and nonbattle injury (DNBI). Verify that all mental health/combat stress control matters are addressed. Validate that there is a combat lifesaver program. Check the availability of emergency dental and eye services.

| Performance Steps |
|---|

**1.** Integrate combat health support functional areas into the MDMP.

**a.** Confirm patient evacuation (ground and air) routes from companies/troops to the aid station.

**b.** Confirm aid station locations.

**c.** Identify possible casualty lines of drift.

**d.** Verify casualty collection point locations.

**e.** Integrate the medical platoon into the battle plan, if applicable.

**f.** Plan for medical support to civilians, when possible.

***Note:*** Coordinate with the forward support medical company (FSMC) for this support. In the contemporary operational environment (COE), providing medical attention to a child may encourage the family members to provide valuable information on insurgent operations.

**2.** Coordinate the force health protection plan with the battalion medical platoon (MEDPLT).

**a.** Medical treatment at level II.

**b.** Evacuation procedures.

**3.** Coordinate preventive medicine program through the MEDPLT to include the following:

| Performance Steps |
|---|

*Note:* Medical personnel oversee all PVTMED activities in the unit. The PVTMED personnel will assist in training company/troop field sanitation teams in aspects of environmental sanitation, limited control of animal reservoirs, and disease vectors.

**a.** Food, water, and arthropod-borne diseases.

**b.** Environmental injuries, such as heat and cold injuries.

**c.** Consultation on the following areas:

**(1)** Environmental sanitation.
**(2)** Epidemiology.
**(3)** Sanitation engineering.
**(4)** Pest management.

**4.** Verify that commanders support field preventive medicine measures.

**a.** Establish field sanitation teams to preserve the unit's health and reduce incidents of DNBI.

**b.** Enforce food and water safety standards.

**c.** Plan and enforce personnel hygiene measures.

**d.** Develop and enforce a unit sleep plan to provide Soldiers with a minimum of 4 hours of uninterrupted sleep in a 24-hour period.

**5.** Ensure that all mental health/combat stress control matters are addressed.

**6.** Validate the combat lifesaver program.

**a.** Each company/troop must have a combat lifesaver program.

**b.** Each squad will have a qualified combat lifesaver.

**c.** Ask the FSMC for training assistance, if needed.

**7.** Coordinate emergency dental service.

**8.** Coordinate emergency eye wear/care with optometry section at the MEDPLT.

**Evaluation Preparation:** *Setup:* Prepare a scenario that requires the Soldier to respond to questions about the performance measures. This may be presented orally or in writing.

*Brief Soldier:* Tell the Soldier that he/she will be required to correctly respond to questions and that there is not a sequence for the performance measures.

| Performance Measures | GO | NO GO |
|---|---|---|
| **1.** Integrated combat health support functional areas into the MDMP process. | ___ | ___ |
| **2.** Coordinated force health protection plan with the battalion MEDPLT. | ___ | ___ |

| Performance Measures | GO | NO GO |
|---|---|---|
| **3.** Coordinated preventive medicine program through the MEDPLT. | ___ | ___ |
| **4.** Verified that commanders support field preventive medicine measures. | ___ | ___ |
| **5.** Ensured that all mental health/combat stress control matters were addressed. | ___ | ___ |
| **6.** Validated the combat lifesaver program. | ___ | ___ |
| **7.** Coordinated emergency dental service. | ___ | ___ |
| **8.** Coordinated emergency eye wear with optometry section at MEDPLT. | ___ | ___ |

**Evaluation Guidance:** Refer to chapter 1, paragraph 1-4 b (6).

**References**

**Required:**

**Related:** FM 21-10, FM 4-02.4 (FM 8-10-4), and FM 8-55

## 081-831-1057

### Supervise Compliance with Preventive Medicine Measures

**Conditions:** You are a unit leader. Your unit is deployed to the field. You have the equipment authorized by your table of organization and equipment (TOE), field sanitation equipment and supplies, and a trained field sanitation team (FST).

***Note:*** Health Care Specialists (68W), organic or attached to deployed units, normally fulfill the requirement as the FST with additional training.

**Standards:** Ensure company-sized units establish and employ manned, trained, and equipped unit FSTs. Ensure that all individual preventive medicine measures (PMMs) are implemented to reduce the threat of disease and nonbattle injury (DNBI) which will assist in mission accomplishment. Ensure that personnel are aware of the major components of the medical threat to field forces and are following the individual PMMs necessary to prevent DNBI. Ensure that selected unit personnel receive FST training on the team's major areas of responsibility.

**Performance Steps**

**1.** Ensure that personnel are aware of the major components of the medical threat to field forces.

- **a.** Heat. Types of heat injuries.
  - **(1)** Heat cramps.
  - **(2)** Heat exhaustion.
  - **(3)** Heatstroke (medical emergency).
- **b.** Cold. Types of cold injuries.
  - **(1)** Chilblain.
  - **(2)** Immersion foot.

**Performance Steps**

(3) Trench foot.

(4) Frostbite.

(5) General hypothermia (medical emergency).

**c.** Arthropods (insects).

(1) Common diseases transmitted directly by arthropods.

**(a)** Mosquitoes—malaria, yellow fever, dengue fever, and encephalitis.

**(b)** Some ticks, as well as mosquitoes—encephalitis.

**(c)** Sand flies—sand fly fever and leishmaniasis.

**(d)** Body lice—epidemic typhus.

**(e)** Hard ticks—Lyme disease.

**(2)** Common diseases transmitted by insects associated with rodents.

**(a)** Fleas—plague and endemic typhus.

**(b)** Mites—scrub typhus.

**d.** Foodborne and waterborne diseases.

**(1)** Common waterborne diseases.

**(a)** Typhoid fever.

**(b)** Cholera.

**(c)** Traveler's diarrhea.

**(d)** Hepatitis A.

**(2)** Common foodborne diseases.

**(a)** Traveler's diarrhea.

**(b)** Cholera.

**(c)** Salmonellosis.

**(d)** Hepatitis.

**e.** Toxic industrial materials (TIMs).

**(1)** Examples of TIMs.

**(a)** Carbon monoxide.

**(b)** Hydrogen chloride.

**(c)** Bore/gun gases.

**(d)** Solvents, greases, insecticides, and oils.

**(2)** Harmful effects of TIMs.

**(a)** Skin irritation.

**(b)** Asphyxiation (choking, suffocation).

**Performance Steps**

(c) Central nervous system depression.

(d) Death.

f. Noise hazards.

(1) Examples of noise hazards.

(a) Weapons.

(b) Aircraft.

(c) Most military vehicles and generators.

(2) Harmful effects of exposure to noise hazards.

(a) Temporary loss of hearing—lasts minutes to hours.

(b) Permanent loss of hearing.

g. Other medical threats to field forces.

(1) Skin disease—common in extremely dry or humid climates.

(2) Altitude sickness—locations above 8,000 feet.

(3) Poisonous plants and animals.

(4) Tobacco use.

(5) Poor medical threat intelligence.

2. Ensure that personnel are following the individual PMMs necessary to prevent DNBI. (See STP 21-1-SMCT, task 081-831-1053.)

a. Protect against cold injuries.

b. Protect against heat injuries.

c. Protect against arthropods and arthropod-borne diseases.

d. Protect against poisonous plants and animals.

e. Protect against diseases from contaminated food and water.

f. Protect against diseases from human waste.

g. Protect against diseases from exposure to soil and common objects.

h. Maintain personal hygiene.

i. Maintain the proper level of nutrition.

j. Take measures to resist stress.

k. Protect against sexually transmitted diseases.

l. Protect against acquired immune deficiency syndrome (AIDS).

m. Avoid adverse effects of tobacco products.

| Performance Steps |
|---|

**3.** Ensure that FST personnel assist the commander in reducing DNBI during all deployment phases.

*Note:* Training of FST members enables unit commanders to provide limited control of insects, proper disinfection of water, and safe food supplies. Training of personnel as FST members will be provided by supporting preventive medicine resources.

**a.** Requirements for FST members, when no organic medical personnel are available.

**(1)** Two Soldiers are selected to receive FST training.

**(2)** One Soldier must be a noncommissioned officer (NCO).

**(3)** Selected team members should have at least six months service remaining with their unit.

**(4)** These Soldiers should receive training from preventive medicine (PM) personnel according to AR 40-5.

**b.** FST tasks and/or responsibilities. The unit FST performs the following tasks in the unit area:

**(1)** Inspect water containers and trailers.

**(2)** Disinfect and monitor unit water supplies for chlorine residual.

**(3)** Monitor unit field food service operations.

**(4)** Monitor unit waste disposal operations.

**(5)** Control arthropods, rodents, and other animals in the unit area.

**(6)** Train unit personnel in using individual PMM.

**(7)** Monitor the status of PMM in the unit.

**(8)** Assist in selecting the unit bivouac site.

**(9)** Supervise the construction of field sanitation devices.

**(10)** Monitor unit personnel when applying individual PMM.

**Evaluation Preparation:** *Setup:* Evaluate each Soldier individually during a field training exercise (FTX) or normal training session. Use the location, weather conditions, and duration of the FTX as the scenario to base your evaluation questions around. If the evaluation is conducted during normal training sessions, create a scenario as the basis for your evaluation questions.

*Brief Soldier:* Tell the Soldier that he/she will be evaluated on his/her ability to answer preventive medicine measure questions pertaining to the training scenario provided.

| Performance Measures | GO | NO GO |
|---|---|---|
| **1.** Ensured that personnel were aware of the major components of the medical threat to field forces. | ___ | ___ |
| **2.** Ensured that personnel were following individual PMMs to prevent DNBI. | ___ | ___ |
| **3.** Ensured that FST personnel assisted the commander in reducing DNBI during all deployment phases. | ___ | ___ |

**Evaluation Guidance:** Refer to chapter 1, paragraph 1-4 b (6).

**References**

**Required:**

**Related:** AR 40-5, CHPPM Tobacco, FM 21-10, FM 4-02.17 (FM 4-02.17), FM 4-25.12 (FM 21-10-1), and Hooah4Health

**SUBJECT AREA 3: CHEMICAL, BIOLOGICAL, RADIOLOGICAL, AND NUCLEAR**

## 031-503-1001

### Chemical-Agent Identify Chemical Agents Using an M256A1 Detector Kit

**Conditions:** You are in a chemically contaminated environment. You are given an M256A1 chemical-agent detector kit, your assigned protective mask, mission-oriented protective posture (MOPP) gear, a watch, TM 3-6665-307-10, TM 43-0003-30, and FM 3-11.5 (FM 3-5). This task will be performed in MOPP4.

**Standards:** Identify chemical agents using an M256A1 chemical-agent detector kit by performing operator checks on the kit, putting the kit into operation, using the correct sequence, and identifying the agent(s) within the limitations of the kit without becoming a casualty.

| Performance Steps |
|---|

**1.** Perform before-operation, preventive maintenance checks and services (PMCS) on the M256A1 chemical-agent detector kit.

**a.** Ensure that the expiration date on the kit has not passed.

**b.** Ensure that M8 detector paper is present.

**c.** Ensure that there are at least four sampler-detectors in the kit.

**2.** Prepare the kit for use.

**a.** Read all instruction cards in the kit.

**b.** Remove one sampler-detector from the kit, and read the instructions printed on the bag.

***Note:***Open the sampler-detector bag, and conduct tests while facing into the wind to keep the vapors from your equipment and clothing from contaminating the test results.

***Note:*** Do not expose the sampler-detector to heavy rain or other forms of water because the test results could be tainted.

***Note:*** Do not touch the sampler-detector test spots because dirt or oil from your gloves could cause the test results to be tainted.

**Performance Steps**

**c.** Remove the sampler-detector from the bag (save the bag, so you can use the instructions printed on the outside). Dispose of the sampler-detector if there are broken or missing ampoules, missing spots, or crushed reagent channels, or if the blood-agent test spot is pink.

**3.** Test for toxic-agent vapors.

**a.** Swing out the heater, and remove and save the two heater pads (used for breaking the glass ampoules and holding the heater assembly down). Swing the heater back in. Keep the protective strips over the spots.

**b.** Remove the pull tab marked "1" to expose the lewisite-detecting tablet. Bend the tab marked "2" over the lewisite-detecting tablet, and rub the upper half of the tab until a mark is visible.

**WARNING**

Before breaking the glass ampoules (except the heater ampoules), place one heater pad on each side of the sampler-detector covering the ampoule to be broken. These pads will prevent pieces of glass from cutting your gloves and hands.

**c.** Hold the sampler-detector with the test spots and the arrow pointing up, and crush the four center ampoules marked "3."

***Note***: Nerve-agent test spots may be difficult to wet with the solution as the kit gets older. Work solutions into the spot carefully while pressing the protective strip over the nerve-agent test spot.

**d.** Turn the sampler-detector so that the arrow points down. Using the heater pads, squeeze the ampoules to force the liquid through the formed channels onto the test spots.

**e.** Hold the sampler-detector with the arrow pointing down, and hold your thumb on the protective strip over the middle test spot.

**f.** Swing the heater away from the blister test spot.

**WARNING**

Avoid hot vapors that may burn you when crushing the heater ampoules. Hold the sampler-detector to one side while venting it to avoid hot vapors.

Do not use the heater pads to crush the green heater ampoules marked "4."

**g.** Activate the first heater ampoule marked "4" by crushing one green ampoule.

**Performance Steps**

**CAUTION**

Apply only a slight amount of pressure to hold down the heater assembly to avoid damaging the blister spot.

**WARNING**

Do not touch any of the sampling spots with your glove while holding the heater assembly.

***Note***: Place one of the heater assembly pads on top of the heater assembly while holding it down on the test spot in order to provide an added level of protection.

**h.** Swing the heater immediately back over the test spot, and place a heater assembly pad on top of the heater. Apply gentle pressure to the top of the heater with your thumb while using your index and middle fingers to support the back of the heater assembly.

**i.** Wait 2 minutes, and then swing the heater and protective strip away from the test spots.

**WARNING**

Do not hold the sampler-detector in direct sunlight while exposing the test spots because the test results could be tainted.

**j.** Expose the test spots to air for 10 minutes, but shield them from direct sunlight. (The sampler-detector can be laid down or held by the hinged protective strip.)

**k.** Wait 10 minutes, and then crush the second green ampoule marked "4." Immediately swing the heater back over the blister test spot. Place the heater assembly pad on top of the heater, and apply gentle pressure to the top of the heater with your thumb while using your index and middle fingers to support the back of the heater assembly.

**l.** Wait 1 minute, and then swing the heater away from the test spot.

**m.** Hold the sampler-detector with the arrow pointing down and the test spots exposed. Use the heater pads to crush the remaining ampoules marked "5." Wet the test spots by squeezing the ampoules and forcing the liquid onto the test spots.

**n.** Bend the tab marked "2" over the lewisite-detecting tablet, and rub the bottom half of the tab until a mark is visible.

**o.** Compare the sampler-detector test colors to determine hazard conditions. Turn the sampler-detector upside down, and compare the colors of the test spots (including the lewisite tab) with those shown on the sampler-detector.

**Performance Steps**

*Note:* Blister-agent test spots (H and CX) develop color immediately after the ampoules are broken.

*Note:* Blood-agent and lewisite tests may be compared after 10 minutes of exposure time. A nerve-agent test requires a waiting period of 3 minutes after the ampoules have been broken.

**(1)** Compare the nerve-agent test after a 3-minute wait. If no color develops, a positive nerve test is indicated. Disregard any small, blue or bluish green areas under the plastic rim of the nerve agent spot.

**(2)** Compare the lewisite test after about 10 minutes of exposure time. Look for a change in the color of the rub marks on the lewisite-detecting tab marked "2." Look very closely. At low concentrations, the color change may be very slight.

**(3)** Compare the blood-agent (round spot) test after about 10 minutes of exposure time.

**(a)** Yellow or orange colors sometimes occur when no agent is present.

**(b)** Pink or blue colors must be present to indicate blood agents. Any combination of colors, including a rainbow effect that includes pink or blue, should be considered a positive blood-agent test.

*Note*: If a blood agent is indicated, repeat the testing for toxic-agent vapors with a fresh sampler-detector for blood agents only. If a blood agent is not indicated the second time, then a blood agent is not present. If a blood agent is indicated the second time, then a blood agent is present.

**4.** Report the results to your supervisor.

**5.** Dispose of the expended or unserviceable materials.

**WARNING**

Each sampler-detector contains 2.6 milligrams (mg) of mercuric cyanide and should be considered hazardous waste (HW). Used or expired kits must be demilitarized and turned in to the Defense Reutilization and Marketing Office (DRMO) according to TM 43-0003-30.

**a.** Dispose of the expended or unserviceable materials according to federal, state, and local laws; military regulations and publications; host-nation laws (if more restrictive than United States [U.S.] laws); and local standing operating procedures (SOPs). At a minimum, place the used decontaminating materials in a sealed plastic bag (such as a zipper-type bag) and label the bag with its contents.

**b.** Dispose of contaminated HW materials according to FM 3-11.5 (FM 3-5).

**6.** Perform after-operation PMCS on the M256A1 chemical-agent detector kit.

**a.** Ensure that the expiration date on the kit has not passed.

**b.** Ensure that the M8 detector paper is present.

**c.** Ensure that there are at least four sampler-detectors remaining in the kit.

**Evaluation Preparation:** *Setup*: Provide the Soldier with the items listed in the task conditions statement. Evaluate this task during a field exercise or a tactical training session. Use M256A1 training aid kits for training and evaluation purposes.

*Brief Soldier*: Tell the Soldier to perform operator checks of the chemical-agent detector kit, put the kit into operation using the correct sequence while in MOPP4, and identify the agent(s) within the limitations of the kit.

| Performance Measures | GO | NO GO |
|---|---|---|
| **1.** Performed before-operation PMCS on the M256A1 chemical-agent detector kit. | —— | —— |
| **2.** Prepared the kit for use. | —— | —— |
| **3.** Tested for toxic-agent vapors. | —— | —— |
| **4.** Reported the results to your supervisor. | —— | —— |
| **5.** Disposed of expended or unserviceable materials. | —— | —— |
| **6.** Performed after-operation PMCS on the M256A1 chemical-agent detector kit. | —— | —— |

**Evaluation Guidance:** Refer to chapter 1, paragraph 1-4 b (6).

**References**

**Required:** FM 3-11.5 (FM 3-5)., TM 3-6665-307-10, and TM 43-0003-30

**Related:**

## 031-503-1002

### Conduct Unmasking Procedures

**Conditions:** You are given a tactical situation, a group of Soldiers in mission-oriented protective posture (MOPP) 4, an M256-series chemical-agent detector kit, a chemical-agent monitor (CAM), M8 and M9 detector paper, a watch, Technical Manual (TM) 3-6665-307-10, Field Manual (FM) 3-11.4, an area where chemical agents have been used, and one of the following situations:

1. An M256A1 chemical-agent detector kit is available.
2. An M256A1 chemical-agent detector kit is not available.

**Standards:** Conduct unmasking procedures in sequence with or without using the M256-series detector kit and without incurring casualties.

| Performance Steps |
|---|
| ***Note:*** Before conducting unmasking procedures, make every effort to confirm the absence of chemical contamination. A CAM, an M256-series chemical-agent detector kit, and M8 and M9 detector paper should be used along with a visual check of the area. |
| ***Note:*** The senior person present selects one or two Soldiers to unmask after permission is received from higher headquarters. |
| **1.** Conduct unmasking procedures in the following sequence using an M256-series detector kit. |

**Performance Steps**

*Note:* Conduct unmasking procedures in the shade when possible.

**a.** Use an M256-series detector kit to test for chemical agents. Use M8 detector paper to check for possible liquid contamination. Continue unmasking procedures only if both tests are negative.

**b.** Direct selected Soldiers to unmask for 5 minutes and then don, seal, and clear their masks. If symptoms appear, tell Soldiers to mask immediately, and then treat them for exposure.

**c.** Observe Soldiers for 10 minutes for chemical-agent symptoms.

**d.** Direct all Soldiers to unmask if no symptoms appear.

**e.** Check Soldiers for delayed symptoms. Have first aid treatment available.

**2.** Conduct unmasking procedures in the following sequence without using an M256-series detector kit.

*Note:* Conduct unmasking procedures in the shade when possible.

**a.** Use M8 detector paper and/or a CAM to check for possible liquid contamination. Continue unmasking procedures only if the test is negative.

**b.** Direct selected Soldiers to take a deep breath, break the seals of their masks for 15 seconds (keeping their eyes open), and then seal and clear their masks.

**c.** Observe Soldiers for 10 minutes for chemical-agent symptoms.

**d.** Direct selected Soldiers to break the seals of their masks. If no symptoms appear, direct Soldiers to take two or three breaths, and then seal and clear their masks.

**e.** Observe Soldiers for 10 minutes for chemical-agent symptoms.

**f.** Direct selected Soldiers to unmask for 5 minutes. If no symptoms appear, direct Soldiers to don, seal, and clear their masks.

**g.** Observe Soldiers for 10 minutes for chemical-agent symptoms.

**(1)** If symptoms appear, mask Soldiers immediately and treat for exposure.

**(2)** If no symptoms appear, direct all Soldiers to unmask.

**h.** Check Soldiers for delayed symptoms. Have first aid treatment available.

**3.** Report the absence of contamination in the area and the successful completion of unmasking procedures to higher headquarters.

**Evaluation Preparation:** *Setup:* Evaluate this task during a field exercise or a tactical training session. The M256A1 trainer kit will be used for training and evaluation purposes.

*Brief Soldier:* Tell the Soldier that he/she will be evaluated on his/her ability to conduct unmasking procedures (with or without using a chemical-agent detector kit). The Soldiers participating in the task will act only as directed by the Soldier conducting the unmasking exercise and will not be evaluated on their performance.

| Performance Measures | GO | NO GO |
|---|---|---|
| **1.** Conducted unmasking procedures in the sequence given using an M256A-series detector kit. | — | — |
| **2.** Conducted unmasking procedures in the sequence given without using an M256A1-series detector kit. | — | — |
| **3.** Reported the absence of contamination in the area and the successful completion of unmasking procedures to higher headquarters. | — | — |

**Evaluation Guidance:** Refer to chapter 1, paragraph 1-4 b (6).

**References**

**Required:** FM 3-11.4 and TM 3-6665-307-10

**Related:**

## 031-503-1005

### Submit a Chemical, Biological, Radiological, and Nuclear (CBRN) Report

**Conditions:** You are in an area where a chemical, biological, radiological, and nuclear (CBRN) attack has just occurred. You are given a watch, map, compass, protractor, pencil, paper, FM 3-11.3 (FM 3-3/FM 3-3-1), FM 3-11.4, GTA 03-06-008, and NBC report format guide. You have reacted to the CBRN attack and submitted an initial spot report (SPOTREP). You are required to submit the NBC 1 report to higher headquarters. This task may be performed in mission-oriented protective posture (MOPP) 4.

***Note:*** The NBC report format guide can take many forms (such as the unit tactical standing operating procedure (TSOP) with an NBC 1 report format, or GTA 03-06-008.

**Standards:** Submit the NBC 1 report with the required line information for a nuclear or chemical-biological (CB) attack, and send the report to higher headquarters. Standards are not degraded due to performance in MOPP4.

| Performance Steps |
|---|

**1.** Gather available data to submit the NBC 1 report.

**2.** Fill out the required information as outlined in GTA 03-06-008 or FM 3-11.3 (FM 3-3) (CB) or FM 3-11.3 (FM 3-3-1) (nuclear).

**a.** Line B: Location of observer (use grid coordinates or place name).

**b.** Line D: Date-time group (DTG) of the attack (specify local or Zulu time).

| Performance Steps |
|---|

**c.** Line H: Type and height of burst (if nuclear) or type of agent and persistency (if CB).

**d.** Line C (direction of attack in miles or degrees from the observer), or Line F (location of attack with grid coordinates or place name).

**3.** Select the proper communication precedence.

***Note***: Flash reports should not be delayed because of the lack of information.

**a.** Use flash precedence if this is the first attack of its type (such as the first nuclear, first biological, or first chemical).

***Note***: A flash precedence is used to report the first use of CBRN weapons against United States (U.S.) troops.

**b.** Use immediate precedence for all other attacks.

***Note***: Line L (nuclear) is measured 5 minutes after the attack, and Line M (nuclear) is measured 10 minutes after the attack. Submit the NBC 1 report after information for Line L or M has been gathered. The M256-series chemical-agent detector kit takes 16 minutes to produce reliable results. Submit the NBC 1 chemical report after this test has been done.

**4.** Submit the NBC 1 report.

**a.** All units submit the NBC 1 report to their higher headquarters.

**b.** Units selected by the division level CBRN center as designated observers for nuclear bursts will also submit the NBC 1 report directly to the division level CBRN center.

**Evaluation Preparation**: *Setup*: Provide the Soldier with the items listed in the task conditions statement. Develop a situation containing observer data (this information may be written). A different situation should be developed for each type of report (nuclear or CB).

*Brief Soldier*: Tell the Soldier that the test will consist of preparing and submitting NBC 1 reports.

| Performance Measures | GO | NO GO |
|---|---|---|
| **1.** Gathered available data to submit the NBC 1 report. | —— | —— |
| **2.** Filled out the required information. | —— | —— |
| **3.** Selected the proper communication precedence. | —— | —— |
| **4.** Submitted the NBC 1 report to higher headquarters. | —— | —— |

**Evaluation Guidance:** Refer to chapter 1, paragraph 1-4 b (6).

**References**

**Required:**

**Related:**

## 031-503-1010

### Supervise the Employment of Chemical, Biological, Radiological, and Nuclear (CBRN) Markers

**Conditions:** You are in a tactical environment where CBRN weapons have been used. The contamination has been located and identified. You and your unit are at the appropriate mission-oriented protective posture (MOPP) level. You have CBRN markers, a grease pencil or an CBRN contamination marking set, FM 3-11.3 (FM 3-3), TM 3-9905-001-10, and the requirement to supervise employing CBRN markers.

**Standards:** Supervise employing CBRN markers. Ensure that the appropriate marker is selected, the required information is recorded on the marker, and the marker is properly emplaced. There is no change in standards if this task is performed in MOPP4.

| Performance Steps |
|---|

**1.** Supervise the employment of CBRN markers for nuclear contamination.

**a.** Ensure that all exposed skin is covered (shirt sleeves down, shirt buttoned up, and nose and mouth covered with a rag or t-shirt) prior to actually marking the contaminated area.

**b.** Ensure that markers are placed at the location where a dose rate of 1 centigray per hour (cGyph) or more was measured.

**c.** Ensure that all information is printed on the front side of the marker so that the word "atom" is facing toward you in an upright position. Information should include—

**(1)** The dose rate in cGyph.

**(2)** The date-time group (DTG) (specify local or Zulu) of the detonation. If the DTG is not known, print "unknown."

**(3)** The DTG (specify local or Zulu) of the reading.

**d.** Ensure that markers are positioned so that they can be easily seen and that the recorded information faces away from the area of contamination.

**(1)** Ensure that markers are attached to objects (such as trees or poles) that will be visible from all probable routes through the contaminated area.

**(2)** Ensure that each marker is placed so that the next one can be seen from the last one emplaced.

**2.** Supervise employing CBRN markers for biological contamination.

**a.** Ensure that individuals are at MOPP4 prior to actually marking the contaminated area.

**b.** Ensure that all information is printed on the front side of the marker so that the word "bio" is facing toward you in an upright position. Information should include—

**(1)** The type of agent detected. If unknown, print "unknown."

**(2)** The DTG (specify local or Zulu) of detection.

| Performance Steps |
|---|
| **(3)** The DTG (specify local or Zulu) of the detonation (place beneath the DTG of detection). |
| **c.** Ensure that markers are positioned so that they can be easily seen and that the recorded information faces away from the area of contamination. |
| **(1)** Ensure that markers are attached to objects (such as trees or poles) that will be visible from all probable routes through the contaminated area. |
| **(2)** Ensure that each marker is placed so that the next one can be seen from the last one emplaced. |
| **3.** Supervise employing CBRN markers for chemical contamination. |
| **a.** Ensure that individuals are at MOPP4 prior to actually marking the contaminated area. |
| **b.** Ensure that all information is printed on the front side of the marker so that the word "gas" is facing toward you in an upright position. Information should include— |
| **(1)** The type of agent detected. If unknown, print "unknown." |
| **(2)** The DTG (specify local or Zulu) of detection. |
| **(3)** The DTG (specify local or Zulu) of the detonation (place beneath the DTG of detection). |
| **c.** Ensure that markers are positioned so that they can be easily seen and that the recorded information faces away from the area of contamination. |
| **(1)** Ensure that markers are attached to objects (such as trees or poles) that will be visible from all probable routes through the contaminated area. |
| **(2)** Ensure that each marker is placed so that the next one can be seen from the last one emplaced. |

**Evaluation Preparation:** *Setup*: Use simulated agents to produce a contaminated environment for chemical and biological (CB) agents. When requiring a marker for nuclear contamination, tell the Soldier the amount of radiation present.

*Brief Soldier:* Tell the Soldier that the test will consist of ensuring that CBRN markers are properly emplaced and all information is printed on the markers.

| Performance Measures | GO | NO GO |
|---|---|---|
| **1.** Supervised employing CBRN markers for nuclear contamination. | —— | —— |
| **2.** Supervised employing CBRN markers for biological contamination. | —— | —— |
| **3.** Supervised employing CBRN markers for chemical contamination. | —— | —— |

**Evaluation Guidance:** Refer to chapter 1, paragraph 1-4 b (6).

**References**

**Required:** FM 3-11.3 (FM 3-3) and TM 3-9905-001-10

**Related:**

## 031-503-2053

### Report CBRN Information Using NBC 4 Reports

**Conditions:** You are in a CBRN contaminated tactical environment, given a watch, a map, a compass, a protractor, a pencil, paper, and graphic training aid (GTA) 03-06-008.

**Standards:** Report CBRN information using the Nuclear, Biological, and Chemical (NBC) 4 report. Complete NBC 4 reports with all heading information and mandatory line items (Q, R, and S for nuclear, or I, Q, S, and T for chemical, biological, or release other than attack [ROTA]). Include all other appropriate data, and ensure that each report is in the correct format. Disseminate completed NBC 4 reports to the proper authority.

| Performance Steps |
|---|

***Note:*** A sample copy of a NBC 4 report is located in Field Manual 3-11.3, appendix L.

**1.** Go to step 2 if it is a nuclear report. Go to step 3 if it is a chemical and biological (C/B) report. Go to step 4 if it is a ROTA report.

***Note:*** Depleted uranium (DU) and DU/low level radiation (LLR) materials (DULLRAM) will be treated as a radioactive hazard; proceed to step 2.

**2.** Nuclear report.

***Note:*** The mandatory information in a NBC 4 Nuclear Report is location of the reading (line Q), dose rate (line R), and date-time group (DTG) of the reading (line S). The operationally determined information (should be provided if known/command discretion) is the strike serial number (line A), cater description (line K), sensor information (line W), downwind direction and speed (line Y), actual weather conditions (line Z), and general text (GENTEXT). Indicate whether the information is operationally determined (O) or mandatory (M) in the condition column.

**a.** Report the heading information for the NBC 4 Nuclear Report:

**(1)** From: Enter your unit identification.

**(2)** To: Enter unit identification of the receiving unit.

**(3)** Precedence: Use IMMEDIATE.

**(4)** Security Classification: Usually sent UNCLASSIFIED.

**(5)** DTG Sent: Use eight digits (DDHHMM - 2 digits for the day, four digits to represent military time) plus ZULU or LOCAL.

**(6)** Category of Report: Enter INITIAL if this is the first report on this attack you will submit; otherwise, enter FOLLOW-UP.

**b.** Report attack information.

**(1)** Line A: Enter strike serial number.

**(2)** Line K: Crater description

**(3)** Line Q: Enter location coordinates of reading/sample/detection and type of sample/detection.

**(4)** Line R: Level of contamination, dose rate trend, and decay rate trend.

**(a)** Dose rate trend/decay rates: BACK = Background, DECR = Decreasing, INCR = Increasing, INIT = Initial, SAME = Same, and PEAK = Peak.

**(b)** Relative decay rates: DN = Decay normal, DF = Decay faster than normal, and DS = Decay slower than normal.

**(5)** Line S: Enter DTG of reading or initial detection of contamination.

***Note:*** Lines Q, R, and S are repeatable up to 20 times in order to describe multiple detectors and monitoring or survey points.

**(6)** Line W: Sensor information.

**(a)** Generic alarm results: POS = Positive, NEG = Negative.

**(b)** Nonspecific Potential Harmful Result: POS = Positive, NEG = Negative.

**(b)** Confirmatory test: Y=Conducted, N= Not Conducted.

**(c)** Confidence level of results: LOW = Confidence, MED = Confidence, HIGH = Confidence.

***Note:*** WHISKEY format is prepared for future use. Procedures on how to use it will follow later.

**(7)** Line Y: Downwind direction and speed.

**(8)** Line Z: Actual weather conditions.

**(9)** Line GENTEXT: Free text (Information that adds significant value to the report).

**c.** Go to step 5.

**3.** Chemical or biological report.

***Note:*** The mandatory information in a NBC 4 Chemical or Biological Report is the release information on biological/chemical attacks or ROTA event (line I), the location of reading/sample/detection and type of sample/detection (line Q), DTG of reading or initial detection of contamination (line S), and terrain/topography and vegetation description (line T). The operationally determined information (should be provided if known/command discretion) is the strike serial number (line A), level of contamination, dose rate trend, and decay rate trend if known (line R), sensor information (line W), downwind direction and speed (line Y), actual weather conditions (line Z), and general text (GENTEXT). Indicate whether the information is operationally determined (O) or mandatory (M) in the condition column.

**a.** Report the heading information for the NBC 4 Chemical or Biological Report:

**(1)** From: Enter your unit identification.

**(2)** To: Enter unit identification of receiving unit.

**(3)** Precedence: Use IMMEDIATE.

**(4)** Security Classification: Usually sent UNCLASSIFIED.

**Performance Steps**

**(5)** DTG Sent: Use eight digits (DDHHMM - 2 digits for the day, four digits to represent military time) plus ZULU or LOCAL.

**(6)** Category of Report: Enter INITIAL if this is the first report on this attack you will submit; otherwise, enter FOLLOW-UP.

**b.** Report attack information:

**(1)** Line A: Enter strike serial number.

**(2)** Line I: Release information on biological/chemical attacks or ROTA events. Type of agent/type of burst.

**(a)** Persistent (P), Nonpersistent (NP), Thickened (T), or Unknown (UNK).

**(b)** Chem/Bio: Air, ground, or spray.

**(c)** Type of detection (see GTA 03-06-008).

**(3)** Line Q: Location coordinates of reading/sample/detection and type of sample/detection (see GTA 03-06-008).

**(4)** Line R: Level of contamination, dose rate trend, and decay rate if known.

**(a)** LDXX lethal dose xx = LD1 to LD99.

**(b)** IDXX incapacitating dosage xx = ID1 to ID99.

**(c)** ICTXX incapacitating dosage xx = ICt1 to ICt99.

**(d)** LCTXX lethal dosage xx = LCt1 to LCt99.

**(e)** MCTXX eye-affecting dosage xx (miosis) = MCt1 to MCt99.

**(5)** Line S: DTG of reading or initial detection of contamination.

**(6)** Line T: Terrain/topography and vegetation description.

***Note:*** Lines Q, R, S,and T are repeatable up to 20 times in order to describe multiple detectors and monitoring or survey points.

**(7)** Line W: Sensor information.

**(a)** Generic alarm results: POS = Positive, NEG = Negative.

**(b)** Nonspecific Potential Harmful Result: POS = Positive, NEG = Negative.

**(b)** Confirmatory test: Y=Conducted, N= Not Conducted.

**(c)** Confidence level of results: LOW = Confidence, MED = Confidence, HIGH = Confidence.

**Performance Steps**

***Note:*** The mandatory information in a NBC 4 ROTA is the release information on biological/chemical attacks or ROTA event (line I), the location of reading/sample/detection and type of sample/detection (line Q), the DTG of reading or initial detection of contamination (line S), and terrain/topography and vegetation description (line T). The operationally determined information (should be provided if known/command discretion) is the strike serial number (line A), level of contamination, dose rate trend, and decay rate trend if known (line R), sensor information (line W), downwind direction and speed (line Y), actual weather conditions (line Z), and general text (GENTEXT). Indicate whether the information is operationally determined (O) or mandatory (M) in the condition column.WHISKEY format is prepared for future use. Procedures on how to use it will follow later.

**(8)** Line Y: Downwind direction and speed.

**(9)** Line Z: Actual weather conditions.

**(10)** Line GENTEXT: Free text (information that adds significant value to the report).

**c.** Go to step 5.

**4.** ROTA report.

*Note:* The ROTA report includes ROTA nuclear (RN), ROTA biological (RB), ROTA chemical (RC) and ROTA unknown (RU).

***Note:*** The mandatory information in a NBC 4 ROTA is the release information on biological/chemical attacks or ROTA event (line I), the location of reading/sample/detection and type of sample/detection (line Q), the DTG of reading or initial detection of contamination (line S), and terrain/topography and vegetation description (line T). The operationally determined information (should be provided if known/command discretion) is the strike serial number (line A), level of contamination, dose rate trend, and decay rate trend if known (line R), sensor information (line W), downwind direction and speed (line Y), actual weather conditions (line Z), and general text (GENTEXT). Indicate whether the information is operationally determined (O) or mandatory (M) in the condition column.

**a.** Report the heading information for the NBC 4 ROTA Report:

**(1)** From: Enter your unit identification.

**(2)** To: Enter unit identification of receiving unit.

**(3)** Precedence: Use IMMEDIATE.

**(4)** Security Classification: Usually sent UNCLASSIFIED.

**(5)** DTG Sent: Use eight digits (DDHHMM - 2 digits for the day, four digits to represent military time) plus ZULU or LOCAL.

**(6)** Category of Report: Enter INITIAL if this is the first report on this attack you will submit; otherwise, enter FOLLOW-UP.

**b.** Report attack information:

**(1)** Line A: Enter strike serial number.

**Performance Steps**

(2) Line I: Release information on biological/chemical attacks or ROTA events. Type of agent/type of burst. (UN/NA identification number [4-digit number taken from the Emergency Response Guidebook])

(3) Line Q: Location of reading/sample/detection and type of sample/detection.

(4) Line R: Level of contamination, dose rate trend, and decay rate if known.

(a) LDXX lethal dose xx = LD1 to LD99.

(b) IDXX incapacitating dosage xx = ID1 to ID99.

(c) ICTXX incapacitating dosage xx = ICt1 to ICt99.

(d) ICTXX incapacitating dosage xx = ICt1 to ICt99.

(e) MCTXX eye-affecting dosage xx (miosis) = MCt1 to MCt99.

(5) Line S: DTG of reading or initial detection of contamination.

(6) Line T: Terrain/topography and vegetation description.

(a) Terrain/topography: FLAT = Flat, URBAN = Urban, HILL = Hill, SEA = Sea, VALLEY = Valley, and UNK = Unknown.

(b) Vegetation: BARE = Bare, SCRUB = Scrubby vegetation, WOODS = Wooded terrain, URBAN = Urban, and UNK = Unknown.

***Note:*** Lines Q, R, S, and T are repeatable up to 20 times in order to describe multiple detectors and monitoring or survey points.

(7) Line W: Sensor information.

(a) Generic alarm results: POS = Positive, NEG = Negative.

(b) Confirmatory test: Y=Conducted, N= Not Conducted.

(c) Confidence level of results: LOW = Confidence, MED = Confidence, HIGH = Confidence.

(8) Line Y: Downwind direction and speed.

(9) Line Z: Actual weather conditions.

(10) Line GENTEXT: Free text (Information that adds significant value to the report).

c. Go to step 5.

5. Submit your NBC 4 Report. The method of transmitting the information depends on the tactical situation and mission of the unit. Method will be specified in operation orders, operation plans, or unit standing operating procedures.

---

**Evaluation Preparation:** *Setup:* Gather the items provided in the conditions statement. Develop a situation containing observer data. The information may be written and given to the Soldier. Develop a different situation for each type of report.

*Brief Soldier:* Tell the Soldier that the test will consist of preparing and submitting NBC 4 reports. Give the Soldier the necessary items, including the data that you developed. Tell the Soldier to prepare and submit NBC 4 reports.

| Performance Measures | GO | NO GO |
|---|---|---|
| **1.** Placed required information in the report format the following: | ___ | ___ |
| ***Note:*** Evaluate the Soldier on steps 2 and 4 if preparing a nuclear report. Evaluate the Soldier on steps 3 and 4 if preparing a CB report. | | |
| **a.** Heading. | | |
| **b.** Line I (chem/bio, and ROTA only). | | |
| **c.** Line Q. | | |
| **d.** Line R (nuclear only). | | |
| **e.** Line S. | | |
| **f.** Line T (chem/bio, and ROTA only). | | |
| **2.** Placed other available data in the report. | ___ | ___ |
| **3.** Submitted the completed report. | ___ | ___ |
| **4.** Submitted the NBC 4 report to the units of concern. | ___ | ___ |

**Evaluation Guidance:** Refer to chapter 1, paragraph 1-4 b (6).

**References**

**Required:** DA Form 1971-10-R, FM 3-11.3, and GTA 03-06-008

**Related:** FM 3-11.9, FM 3-11.19, FM 3-11.34, and FM 3-11.86

# 031-504-1061

## Conduct a Mask Fit Test Using the M41 Protection Assessment Test System (PATS)

**Conditions:** You are given an M41 PATS; M17-, M40-, M42- or M45-series protective masks; PATS operator's manual; TC 3-41; DA Form 2404 (*Equipment Inspection and Maintenance Work Sheet*), and any of the following situations:

1. You are directed by the commander to conduct a mask fit test.
2. You have a Soldier(s) who requires an initial issue, an annual verification, or a replacement mask issue.

***Note:*** This task will not be performed in mission-oriented protective posture (MOPP) 4.

**Standards:** Conduct a mask fit test. Perform all the steps in sequence to verify the fit of a protective mask to an individual's face.

| Performance Steps |
|---|
| **1.** Prepare the M41 PATS for operation. |
| **a.** Unpack all equipment without causing damage. |
| **b.** Inventory the equipment using the operator's manual, and note any missing equipment on DA Form 2404. |

**Performance Steps**

**c.** Perform setup procedures.

**(1)** Attach the tube marked A to the ambient port and the tube marked S to the sample port.

**(2)** Connect the alternating current (AC) power supply, or install the battery.

**(3)** Install the alcohol cartridge.

**(4)** Add alcohol.

**d.** Perform a systems check.

**(1)** Prepare the protection assessment test instrument (PATI).

**(2)** Conduct troubleshooting on the PATS.

**2.** Prepare the mask for the fit test (M40, M42, or M45 series).

**a.** Attach the drink tube sampling adapter to the drink tube quick-disconnect coupling.

**b.** Have the Soldier sit down for the fit test.

**c.** Have the Soldier don the mask.

**d.** Instruct the Soldier to blow as hard as possible several times into the internal drink tube mouthpiece to remove any trapped fluids or foreign matter. Ensure that the drink tube is clear so that foreign matter will not be drawn into the PATI. Ensure that the PATI will be able to draw air from inside the mask. This is a critical step.

**e.** Have the Soldier adjust the face piece and tighten the head harness. Ensure that the face piece is properly fitted.

**3.** Prepare the mask for the fit test (M17 series).

**a.** Attach the drink tube sampling adapter to the drink tube quick-disconnect coupling.

**b.** Attach the drink valve retaining lever to hold the drink valve open.

**c.** Have the Soldier sit down for the fit test.

**d.** Have the Soldier don the mask.

**e.** Instruct the Soldier to blow as hard as possible several times into the internal drink tube mouthpiece to remove any trapped fluids or foreign matter. Ensure that the drink tube is clear so that foreign matter will not be drawn into the PATI. Ensure that the PATI will be able to draw air from inside the mask. This is a critical step.

**f.** Have the Soldier remove the mask. Insert the sample tube extension into the internal drinking tube mouthpiece.

**Performance Steps**

**g.** Remove the drink valve retaining lever from the drink valve lever.

> **WARNING**
>
> Have the Soldier close his/her eyes while putting on the mask to prevent the sample extension tube from injuring his/her eyes.

**h.** Instruct the Soldier to don the mask. Ensure that the sample extension tube is properly positioned.

**i.** Adjust the face piece, and tighten the head harness. Ensure that the face piece is properly fitted.

**j.** Attach the drink valve retaining lever so that the drink valve is held in the OPEN position for the duration of the fit test.

**4.** Conduct a mask fit test. (See the PATS operator's manual.)

***Note:*** Smoking is NOT permitted in the immediate area where the fit test is being conducted and the Soldier should not smoke for at least 30 minutes before the test. To achieve proper results, the Soldier should not talk during the test.

**a.** Remove the high-efficiency particulate air (HEPA) filter from the twin tube assembly.

**b.** Attach the twin sample tube (marked SAMPLE) to the sample port on the end of the drink tube sampling adapter.

**c.** Monitor the reading on the display with the PATI in the count mode and the Soldier remaining still until an acceptable reading of 3.0 particles/cubic centimeters (cm) or lower is obtained.

**d.** Press the FIT TEST key on the keypad after obtaining an acceptable seal to bring the instrument into the standby fit test mode.

***Note:*** Before continuing to the next step, brief the Soldier on how to perform the exercises outlined in the operator's manual. Instruct the Soldier to breathe normally during all exercises except the deep breathing exercise.

**e.** Verify that the number of exercises is set to S by pressing the NUMBER OF EXERCISES key.

**f.** Press the START/STOP TEST key to begin the fit test. Instruct the Soldier to perform the first exercise.

**g.** Have the Soldier remove his/her mask after the test is complete and PASS is indicated.

**h.** Remove all testing attachments from the Soldier's mask.

**i.** Repeat the test if FAIL is indicated. (Repeat steps 3f and 3g and 4a through 4i.)

**Evaluation Preparation:** *Setup*: Provide the Soldier with the items listed in the task conditions statement.

*Brief Soldier:* Tell the Soldier to conduct the mask fit test based on the type of protective mask issued.

***Note:*** Performance measures will vary depending on the type of mask being fitted (M17, M40, M42, or M45 series).

| Performance Measures | GO | NO GO |
|---|---|---|
| **1.** Prepared the M41 PATS for operation. | —— | —— |
| **2.** Prepared the mask for the fit test (M40, M42, or M45 series). | —— | —— |
| **3.** Prepared the mask for the fit test (M17 series). | —— | —— |
| **4.** Conducted a mask fit test. | —— | —— |

**Evaluation Guidance:** Refer to chapter 1, paragraph 1-4 b (6).

**References**

**Required:** TC 3-41

**Related:** DA Form 2401, TM 3-4240-279-10, and TM 3-4240-279-20&P

**SUBJECT AREA 4: SURVIVE (COMBAT TECHNIQUES)**

## 061-283-6003

### Adjust Indirect Fire

**Conditions:** Given a pair of binoculars, a radio, a compass, pencils, a coordinate scale, a map of the target area, a target to engage within the area, and grid location of friendly troops.

**Standards:** Determine the target location to within 250 meters of its actual location. Transmit the initial call for fire within 3 minutes after identifying the target. Send adjustments within 45 seconds after each round impacts. Enter the fire-for-effect phase using no more than six rounds (initial round plus five for adjustment). Fire for effect within 50 meters of the target using successive bracketing procedures (or creeping fire if danger close).

**Performance Steps**

**1.** Locate the target within 250 meters of the actual target location.

**a.** Locate the target by grid coordinates.

**b.** Determine the direction from your position to the target.

**c.** Formulate a call for fire. Include the elements of the call for fire in sequence.

**(1)** Observer identification (your call sign).

**(2)** Warning order (adjust fire).

**(3)** Location of target (grid data).

**(4)** Description of the target (for example "Infantry platoon in the open.").

**Performance Steps**

**(5)** Method of engagement (may be omitted if area fire is desired).

**(a)** If the target is within 600 meters of friendly troops, announce "Danger close" to the fire direction center (FDC) in the initial call for fire in the method of engagement phase.

**(b)** Use creeping procedures to adjust danger close fire. Range corrections should not exceed 100 meters.

**(c)** Report the initial target location on the enemy side of the target.

**(6)** Method of fire control. The request for a fire mission would be similar to figure 061-283-6003-1.

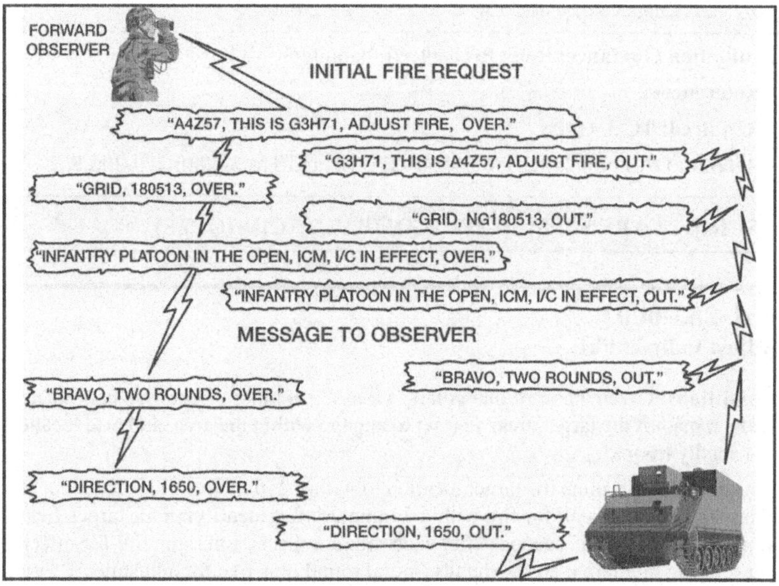


**Figure 061-283-6003-1. Initial fire request**

**2.** Transmit the call for fire to the FDC within 3 minutes of target identification.

**a.** Conduct three transmissions.

**(1)** Send observer identification and warning order. Example: "A4Z57, This is G3H71, Adjust fire, Over"

**(2)** Send target location. Example: "Grid 180513, Over". (Give the six-digit grid of the target, with the grid zone identifier, to within 250 meters of the actual target location.)

**(3)** Send target description, method of engagement, and method of fire and control. Example: "Infantry platoon in the open, ICM in effect, Over."

**Performance Steps**

**b.** Give the direction to the target within 100 mils (M2 compass), or five degrees (lensatic compass), or give an accurate cardinal direction (no compass available) of the target's actual location. This should be sent before the first correction, or with the first correction.

**3.** Adjust rounds to within 50 meters of the target, within 45 seconds of the impact of each adjusting round.

**a.** Spot each round when it impacts as right or left, over or short of your target.

**b.** Determine corrections for deviation left or right of the target.

***Note***: Measure deviation. Measure the horizontal angle in mils, using the reticle pattern in the binoculars or hand measurement of angular deviation. Estimate the range to the target and divide by 1,000. This is the observer-target (OT) factor. If the OT distance is 1,000 meters or greater, the OT factor is expressed to the nearest whole number. If the OT distance is less than 1,000 meters, the OT factor is expressed to the nearest 1/10th. For example, 800 = 0.8. Multiplying the OT factor by the deviation measured in mils produces deviation corrections in meters.

**c.** When the first range spotting is observed, make a range correction that would result in a range spotting in the opposite direction. For example, if the first round is short, add enough to get an over on the next round. This is called successive bracketing (figure 061-283-6003-2). Figure 061-283-6003-3 shows the impact of your initial round. The target is 2,100 meters away. Since the round is beyond the target, you must drop. You estimate that the round is 250 meters beyond the target. Therefore, you must drop 400 meters to start successive bracketing procedures. The round impacted 50 mils left of the target. With an OT factor of 2, the round impacted 100 meters left. Your correction to the FDC is "Right 100, Drop 400, Over."

**WARNING**

DO NOT BRACKET when DANGER CLOSE—it could result in friendly casualties; use the creeping fire procedure (all corrections are 100 meters or less).

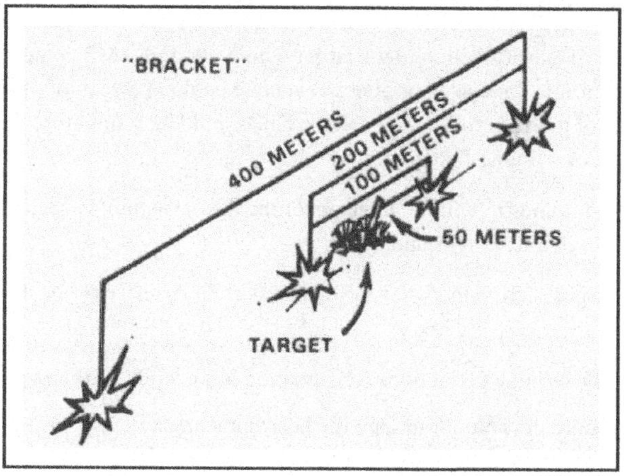


**Figure 061-283-6003-2. Successive bracketing**

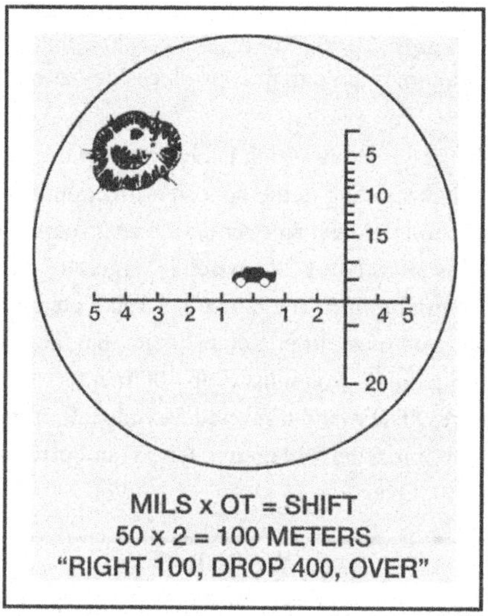


**Figure 061-283-6003-3. Reticle pattern**

**Performance Steps**

**d.** Continue splitting the range bracket until a 100-meter bracket is split or range correct spotting is observed, maintaining deviation on line. (Figure 061-283-6003-4 and figure 061-283-6003-5 show the next adjustments.)

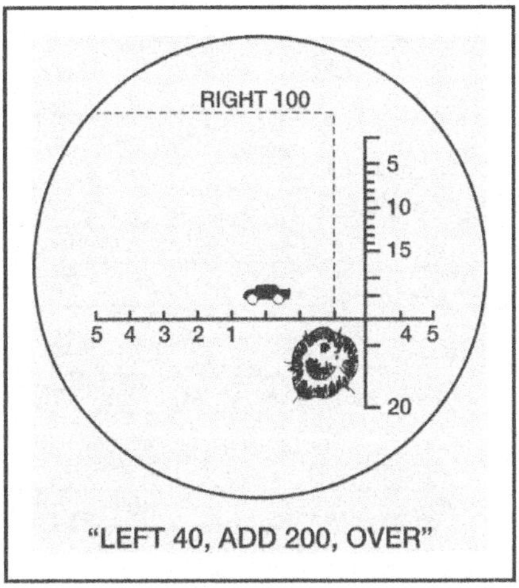


**Figure 061-283-6003-4. Second round**

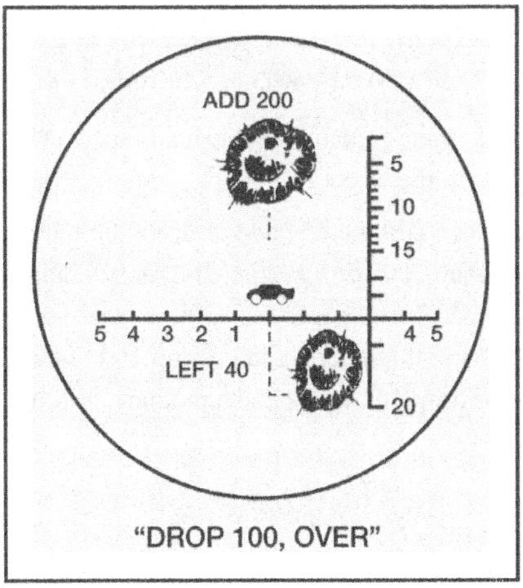


**Figure 061-283-6003-5. Third round**

Performance Steps

**e.** Transmit corrections to the FDC in meters. The initial correction should bracket the target in range. The adjustment phase of a fire mission would resemble the example shown in figure 061-283-6003-6. Deviation correction should be made to keep the rounds on the observer target line.

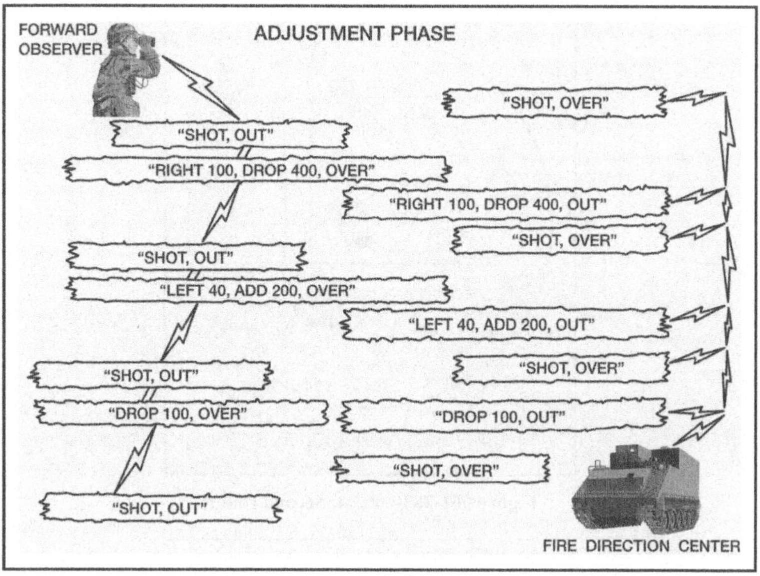


**Figure 061-283-6003-6. Adjustment phase**

**f.** Use the following guide to establish a bracket. When the estimated round impact distance to the target is—

- **(1)** More than 400 meters, add or drop 800 meters.
- **(2)** More than 200 but less than 400 meters, add or drop 400 meters.
- **(3)** More than 100 but less than 200 meters, add or drop 200 meters.
- **(4)** Less than 100 meters, add or drop 100 meters.
- **(5)** Add or drop 50 meters and announce "Fire for effect."

**Performance Steps**

**4.** Initiate fire for effect. When a 100-meter bracket is split or a range correct spotting is made, the fire-for-effect phase is entered (figure 061-283-6003-7). Figure 061-283-6003-8 shows a simulated pattern that might be observed in the fire-for-effect phase and the observed results of fire for effect are reported.

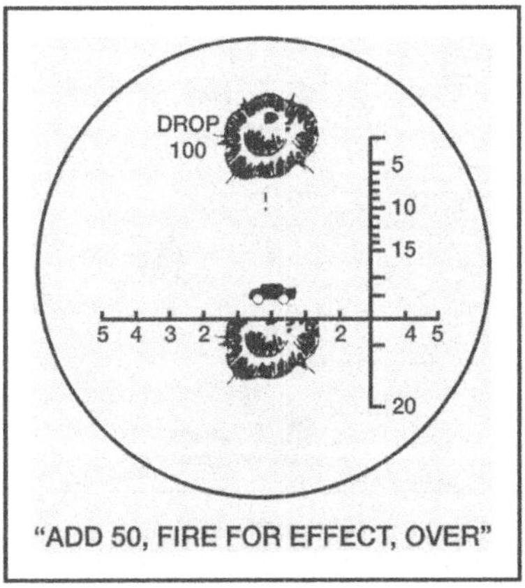


**Figure 061-283-6003-7. Fourth round**

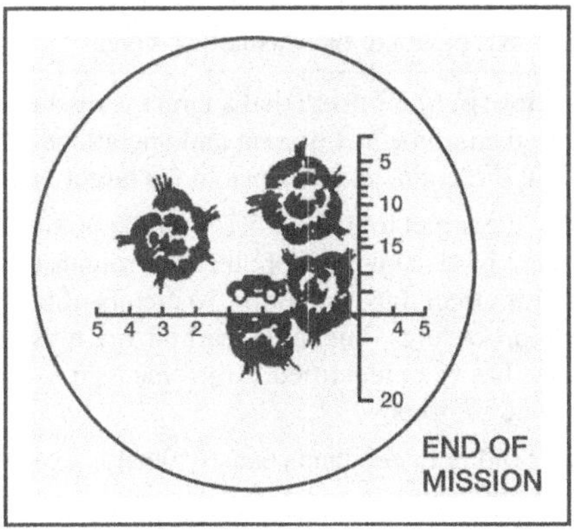


**Figure 061-283-6003-8. Fire-for-effect pattern**

| Performance Steps |
|---|

**5.** Observe the results of fire for effect, transmit refinements (if necessary), and provide end of mission and surveillance (figure 061-283-6003-9).

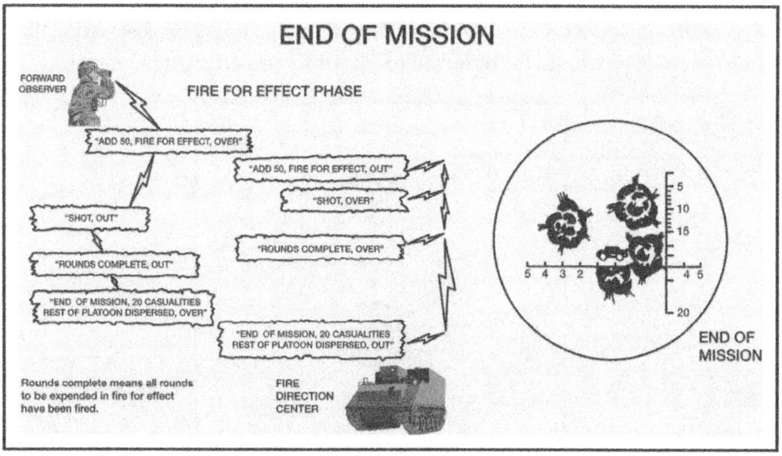


**Figure 061-283-6003-9. End of mission**

**a.** Determine the effects on the target.

**b.** Give a brief description of what happened to the target. Example: "EOM, target destroyed, Estimate two casualties, Over."

**Evaluation Preparation:** *Setup*: Ensure that a target is readily identifiable to the Soldier and the required equipment is present and operational. The evaluator must know the grid location, direction, and distance to the target that will be used.

*Brief Soldier:* Identify the target to the Soldier. Tell the Soldier that he/she must enter the fire-for-effect phase using no more than six rounds (initial round plus five for adjustment). Fire for effect must be within 50 meters of the target, using successive bracketing procedures. The initial call for fire must be made within 3 minutes after the target has been identified. Adjustments must be sent within 45 seconds after each round impacts.

***Note***: Ensure that the Soldier understands exactly what is expected of him/her, but do not help him/her in any way.

| Performance Measures | GO | NO GO |
|---|---|---|
| **1.** Located the target to within 250 meters of the actual target location. | ___ | ___ |
| **2.** Transmitted the call for fire to the FDC within 3 minutes of target identification. | ___ | ___ |
| **3.** Adjusted rounds to within 50 meters of the target, within 45 seconds of the impact of each adjusting round. | ___ | ___ |

| Performance Measures | GO | NO GO |
|---|---|---|
| **4.** Entered the fire for effect phase when a 100 meter bracket was split, or when a range correct spotting was obtained. | —— | —— |
| **5.** Observed the results of fire for effect, transmitted refinements (if necessary), and provided end of mission and surveillance. | —— | —— |

**Evaluation Guidance:** Refer to chapter 1, paragraph 1-4 b (6).

**References**

**Required:**

**Related:** FM 3-25.26 and FM 6-30

---

## 071-326-5705

### Establish an Observation Post

**Conditions:** Given a squad- or platoon-sized element in a defensive position, a TA-312/PT field telephone and communication wire or a radio, and a probable enemy avenue of approach.

**Standards:** Select a location for an observation post (OP) that provides observation of the avenues of approach, is within small arms range of the element, and offers adequate cover and concealment. Establish communications between the OP and the platoon leader or squad leader.

| Performance Steps |
|---|

*Note*: The OPs are generally established along probable avenues of approach to listen and observe, and to provide early warning of enemy approach.

**1.** Select an OP.

**a.** Ensure that the site selected for an OP provides—

**(1)** Maximum observation of the desired area (specified by the platoon leader).

**(2)** Cover and concealment for the occupants of the OP.

**(3)** Concealed routes to and from the OP.

*Note*: Observation is the best way to determine whether the above conditions exist at a site.

**b.** Locate the OP on or near the military crest of a hill, if possible. Avoid topographical crests because of the chance of being skylined. When observation is restricted by the terrain, establishing the OP well down the forward slope might be appropriate (figure 071-326-5705-1).

**Figure 071-326-5705-1. OP position**

**c.** Establish the OPs within effective small arms range of the unit establishing the OP. Ensure the OPs are supported by other supporting fire when possible.

**2.** Establish and operate an OP.

**a.** Wire is the primary means of communication with an OP. You can supplement it with radio. Carefully position and camouflage wire and radio antennas to avoid detection by the enemy (figure 071-326-5705-2).

**Figure 071-326-5705-2. Camouflage of communication lines**

**b.** Personnel going to and from the OP must move carefully to avoid revealing their location to the enemy. Establish separate routes to and from the OP. As camouflage is important, camouflage the OP even when natural concealment is adequate.

**c.** The OPs are operated in reliefs. A minimum of two Soldiers is necessary for each relief. One observes, while the other records and reports observed information. The observer and recorder should switch duties every 20 to 30 minutes because the visual efficiency of an observer decreases rapidly after that length of time.

| Performance Steps |
|---|
| **3.** Establish and operate an OP during limited visibility. |
| **a.** The enemy may use a different, more open avenue of approach during limited visibility conditions; therefore, you might have to move the OP to another position at night. |
| **b.** Limited visibility OPs are usually closer to defensive positions. You can use night vision devices on the OP. The enemy deploys infiltrators against the defense at night. Setting up a series of OPs, backed up by alert troops equipped with night vision devices and snipers, can counter this infiltration. |
| **c.** Operate the OPs in relief, except when movement to and from positions would reveal their locations or endanger personnel. |

**Evaluation Preparation:** *Setup*: In a field environment, with terrain where an OP can be established, provide the element leader with a TA-312/PT field telephone and communication wire or a radio, and two personnel to act as the observer and the recorder.

*Brief Soldier:* Tell the Soldier to establish an OP to observe an area (designated by the tester) forward or to the flanks of the Soldier's element.

| Performance Measures | GO | NO GO |
|---|---|---|
| **1.** Selected an area within small arms range of the squad- or platoon-sized element. | —— | —— |
| **2.** Selected a site that provided maximum observation of the desired area. | —— | —— |
| **3.** Established and operated an OP in area that offered cover and concealment. | —— | —— |
| **4.** Ensured that wire or radio communication was established. | —— | —— |
| **5.** Ensured that radio antenna, if used, was camouflaged. | —— | —— |
| **6.** Established several concealed routes to and from the OP. | —— | —— |
| **7.** Ensured that the OP was camouflaged. | —— | —— |
| **8.** Placed a minimum of two personnel on the OP. | —— | —— |
| **9.** Instructed the observer and recorder to switch duties every 20 to 30 minutes. | —— | —— |

**Evaluation Guidance:** Refer to chapter 1, paragraph 1-4 b (6).

**References**

**Required:** FM 3-21.8 (FM 7-8) and FM 3-21.75 (FM 21-75)

**Related:**

## 071-410-0019

### Control Organic Fires

**Conditions:** Given a unit equipped with table of organization and equipment (TOE) weapons, attached fire support elements (machine gun teams; antitank teams; tube launched, optically tracked, wire guided [TOW] squad; Bradley fighting vehicle [BFV]), an area of responsibility, and the requirement to regulate the first of all weapon systems assigned or attached to your unit.

**Standards:** Assign sectors of fire for each individual and crew-served weapon; issue a priority of target engagement appropriate for each weapon system; inform all personnel of current rules of engagement (ROE); and implement procedures to engage threat targets in your area of responsibility in a timely manner, with the appropriate weapon system, and without causing injury or death to friendly personnel.

**Performance Steps**

> **WARNING**
>
> Soldiers may be killed or injured when firing weapons. Minimum firing distances, backblast areas, and weapon-specific regulations must be strictly followed.

**1.** Implement procedures to apply the following principles of fire control:

**a.** Avoid target overkill; conserve ammunition.

**b.** Use each weapon in its intended role.

**c.** Engage targets that offer a high probability of a kill.

**d.** Engage long-range targets first.

**e.** Destroy targets that pose the greatest threat first.

**2.** Use fire control methods to—

**a.** Maximize the effects of the weapons on the target.

**b.** Achieve mutual support.

**c.** Cover assigned area of responsibility.

**d.** Ensure Soldiers' safety from friendly fire.

**3.** Use the following methods to coordinate and regulate fire:

**a.** Assign sectors of fire. Assign an area for each crew-served weapon, team, or unit to cover by fire. Targets in each assigned sector are the responsibility of the individual(s) assigned to cover the area with fire.

**b.** Assign priority of fires. Tell each Soldier, according to weapon assignment, what to fire at, when, and why. This ensures that each weapon is used in the role it is best suited for. (For example, Dragon gunners engage light armored vehicles; BFV gunners engage tanks with the TOW and BMPs with the 25 mm; riflemen engage dismounted personnel out to 400 meters.)

**Performance Steps**

**c.** Use target reference points (TRPs). Designate recognizable points on the ground (natural or manmade) to use as reference points when identifying sectors of fire or targets, or to control supporting fires.

**d.** Assign final protective fire areas. Assign final protective lines (FPLs) or principal directions of fire (PDF) for machine guns and other automatic crew-served weapons, or ground- or carrier-mounted (MK19, BFV coaxial, 25 mm, and so on).

**4.** Use any of the following techniques to communicate when to start, shift, or cease fire:

**a.** Prearranged signals. Use visual or sound signals (pyrotechnics, whistles, horns, detonation of a device, firing of a weapon) or set a specific time.

**b.** Arm-and-hand signals. Use standard signals when feasible. Remember, personnel must see arm-and-hand signals to respond.

**c.** Fire by example. Initiate firing at or in the direction of the intended target. Have the unit or specific weapon system follow your example.

**d.** Fire commands. Some elements of fire commands may be omitted. If any element is omitted, ensure that the unit or crew thoroughly understands the command. Every fire command should contain the target description and execution element as a minimum. (See table 071-410-0019-1 and table 071-410-0019-2 for examples of fire command elements.)

## Performance Steps

**Table 071-410-0019-1. Examples of fire command elements**

| ELEMENTS | TOW SQUAD | ANTITANK TEAM | CREW-SERVED WEAPON | UNITS |
|---|---|---|---|---|
| 1. **ALERT** | FIRE MISSION or SQUAD | FIRE MISSION GUN #1 | FIRE MISSION GUN #3 | FIRE MISSION, A TEAM, 1ST SQUAD, ETC. |
| 2. **DESCRIPTION** (Direction optional) | TANK | PC | TRUCK WITH TROOPS | TROOPS |
| 2A. **DIRECTION** (if used) | FRONT, 11 O'CLOCK (Clock method) | RIGHT FRONT, 2 O'CLOCK (Clock method) | LEFT OF LONE PINE TREE (TRP method) | CENTER OF SECTOR |
| 3. **RANGE** (in meters) | 1,000 | 600 | 400 | 200 |
| 4. **METHOD OF FIRE** | FRONTAL, DEPTH, OR CROSSFIRE (SITUATION DEPENDENT) | SINGLE, PAIR, VOLLEY, OR SEQUENCE (SITUATION DEPENDENT) | SUSTAINED, SEARCH, OR TRAVERSE, OR COMBINATION (SITUATION DEPENDENT) | SUSTAINED, OR UNTIL TARGET IS DEFEATED |
| 4A. **TYPE MISSILE** | TOW 2 TOW 2A | DRAGON, AT-4 | | |
| 5. **EXECUTION** | FIRE, OR AT MY COMMAND | FIRE, OR AT MY COMMAND | FIRE, OR AT MY COMMAND | FIRE, OR AT MY COMMAND |
| 6. **CLOSING** | CEASE TRACK-ING or CEASE TRACK-ING, OUT OF ACTION | CEASE TRACKING or CEASE TRACKING, OUT OF ACTION | CEASE FIRE | CEASE FIRE |

Table 071-410-0019-2. Examples of fire command elements for a BFV

| ELEMENTS | GUNNERY TECHNIQUES | |
|---|---|---|
| | BATTLESIGHT | PRECISION |
| 1. ALERT | GUNNER | GUNNER |
| 2. WEAPON/ AMMUNITION | BATTLESIGHT | SABOT<br>HE<br>COAX<br>MISSILE |
| 3. DESCRIPTION | TANK, CHOPPER, TRUCK, TROOPS, PC | PC, TRUCK, TROOPS, TANK |
| 4. DIRECTION (Optional) | SHIFT RIGHT (LEFT), TRP-1 (TRP method) or 2 O'CLOCK (clock method) | SHIFT RIGHT (LEFT), TRP-1 (TRP method) or 2 O'CLOCK (clock method) |
| 5. RANGE (in meters)<br><br>GUNNER ID* | <br><br>IDENTIFIED | 600<br><br>IDENTIFIED |
| 6. EXECUTION<br><br>GUNNER'S RESPONSE* | FIRE, OR AT MY COMMAND<br><br>ON THE WAY | FIRE, OR AT MY COMMAND<br><br>ON THE WAY |
| 7. CLOSING | CEASE FIRE | CEASE FIRE |
| *If the BFV gunner cannot identify the target, the Bradley commander (BC) engages the target from his position. The BC announces "From my position, on the way." "Fire" is not announced. | | |

**(1)** Use fire commands to direct the fires of units or key weapons and to place a specific type of fire on certain targets. Fire commands for some weapon systems may vary in form because of weapon characteristics. They should identify who is to fire the direction of fire, type of ammunition, type of target, range, and when to fire.

**(2)** Correct errors in fire commands by stating CORRECTION, correcting the element in error, and then repeating all elements following the corrected element.

**(3)** Use subsequent fire commands to adjust for range or new targets. For example, shift left (or right) up (or down) and designate new target. For subsequent fire commands for a Bradley, use shift left (or right) PC and up (or down) half or full target frame.

**(4)** Develop standing operating procedures (SOPs) and drills for certain actions and commands to make fire control more effective.

| Performance Steps |
|---|

> **CAUTION**
> Dispose of all belt links and spent brass according to the unit SOP.

**Evaluation Preparation:** *Setup*: This task should be evaluated during a field training exercise on a live-fire range. Otherwise, assign a defensive position to a fire team and designate team individual sectors of fire. Provide appropriate ammunition for the weapons assigned. Target areas should represent various types of targets, such as an enemy column formation, a line formation, or a linear target with depth.

*Brief Soldier*: Tell the Soldier to have the team engage the various targets and to use the appropriate weapons on the targets.

| Performance Measures | GO | NO GO |
|---|---|---|
| 1. Assigned individual sectors of fire. | —— | —— |
| 2. Issued a fire command to engage a target. | —— | —— |
| 3. Issued subsequent fire commands as necessary. | —— | —— |
| 4. Gave arm-and-hand signals to control fires. | —— | —— |
| 5. Engaged a target according to the unit SOP. (Graded only if the unit has an SOP for team/squad fire.) | —— | —— |
| 6. Used appropriate weapon(s) on target. | —— | —— |

**Evaluation Guidance:** Refer to chapter 1, paragraph 1-4 b (6).

**References**

**Required:**

**Related:** AR 385-63, FM 3-21.75 (FM 21-75), FM 3-11.3 (FM 3-22/FM 3-3), and FM 3-22.68

## SUBJECT AREA 5: NAVIGATE

## 071-326-0515

### Select a Movement Route Using a Map

**Conditions:** Given an operation or fragmentary order, a 1:50, 000 scale military map, and a compass.

**Standards:** Select a route with the following characteristics:

1. Take advantage of maximum cover and concealment.
2. Ensure observation and fields of fire for overwatch or fire support elements.
3. Allow positive control of all elements.
4. Accomplish the mission quickly, without unnecessary or prolonged exposure to enemy fire.

**Performance Steps**

**1.** Select the route that makes the best use of terrain. Use terrain to your best advantage. Take advantage of the terrain to—

**a.** Cover and conceal the unit during movement.

**b.** Provide maximum effectiveness of the platoon's weapons.

**2.** Use the military aspects of terrain and apply them to any given situation, whether it is a defense, a delay, or a road march behind the forward edge of the battle area (FEBA). Consider the following essential elements:

**a.** Use cover and concealment for any type of movement on the battlefield.

**(1)** Use cover to shield the unit from the effects of weapon fire, especially direct fire. Take advantage of every ravine or depression in the ground to protect and cover the force, especially if forward of the FEBA. Evaluate the terrain, the capabilities of the enemy's weapons systems, and the position of known or suspected enemy emplacements. Visualize a cross section of the terrain and determine where the enemy cannot place effective direct fire on your proposed route.

**(2)** Conceal or disguise the unit. Consider concealment from both air and ground observation. Consider that exhaust smoke or dust can reveal your unit to the enemy.

**b.** Ensure that your proposed route can be covered by fire from overwatch or fire support positions when moving in an area where contact with the enemy is expected.

**(1)** Consider that direct fire weapons must have good observation to fire at known or suspected enemy positions along the movement route. Control the maneuver of your elements, if they make contact. Consider the effects of smoke and dust from friendly and enemy fire.

**(2)** Select a route that gives your unit the best field of fire. They must be in a position to provide suppressive fires immediately. Use crew-served weapons to overwatch the movement. Ensure the overwatch element is able to observe the moving element and provide fire support all the way to the objective. Select overwatch positions that have unobstructed fields of fire to the next overwatch position.

**3.** Ensure that the route can be covered by fire from overwatch or fire support positions. Select the route that provides the most favorable tactical advantage and meets the mission requirements. If enemy air is active or enemy ground forces are in the area of the route, take maximum advantage of cover and concealment. If speed of movement is critical, select the route with the most easily negotiable terrain, avoiding difficult obstacles. Choose the route that provides movement from one easily distinguishable terrain feature to another. Check the terrain based on the above considerations, and then select the quickest and safest route.

| Performance Steps |
|---|
| **4.** Use special purpose maps and aerial photographs when a planning a route. If those aids are available, they provide the most current information. |
| **5.** Reconnoiter the route if time is available and the tactical situation permits it. |

**Evaluation Preparation:** *Setup*: In a field environment, provide the Soldier with a 1:50,000-scale military map of the area and a compass, and issue the Soldier an oral or written operation order.

*Brief Soldier:* Tell the Soldier to select a route of movement between two given points (marked on the map) where the likelihood of enemy contact is unknown. The Soldier must select a route that offers the best cover and concealment, ensure the best observation and fields of fire for support elements, allow positive control of elements, and accomplish the mission without unnecessary or prolonged exposure to enemy fire.

| Performance Measures | GO | NO GO |
|---|---|---|
| **1.** Conducted a map reconnaissance of the area that you had to move over. | —— | —— |
| **2.** Selected a route with the appropriate characteristics. | —— | —— |

**Evaluation Guidance:** Refer to chapter 1, paragraph 1-4 b (6).

**References**

**Required:** FM 7-7 and FM 3-21.8 (FM 7-8)

**Related:**

## 071-329-1019

### Use a Map Overlay

**Conditions:** Given a military map and a company level map overlay.

**Standards:** Position overlay correctly on the map. Identify all graphic symbols and information shown on the map overlay.

| Performance Steps |
|---|
| **1.** Obtain the map sheet(s) listed in the margin. |
| **2.** Locate the grid intersections on the map that correspond to the grid register marks in the opposite corners of the overlay. |
| **3.** Place the overlay on the map so that the grid register marks fall exactly on top of the grid intersections (figure 071-329-1019-1). |

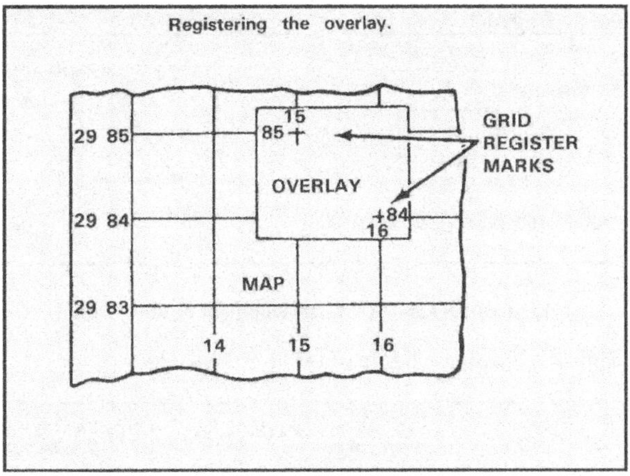


**Figure 071-329-1019-1. Registering the overlay**

**4.** Locate the following points and areas identified on the overlay:

**a.** Basic symbols (figure 071-329-1019-2).

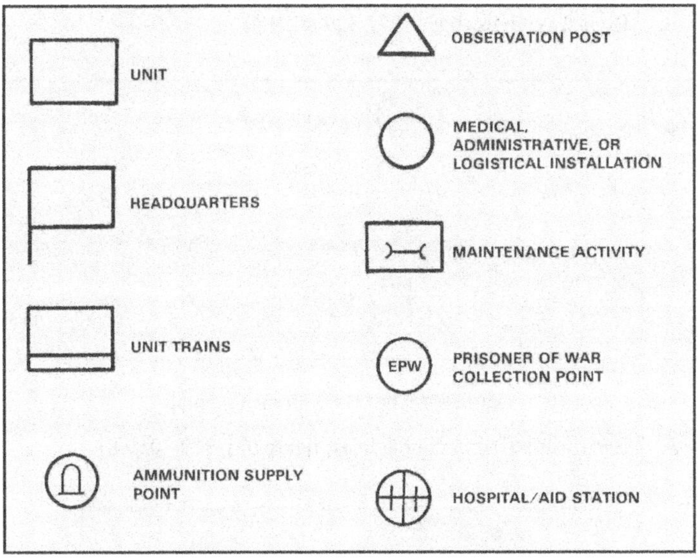


**Figure 071-329-1019-2. Basic symbols**

**b.** Unit symbol (figure 071-329-1019-3).

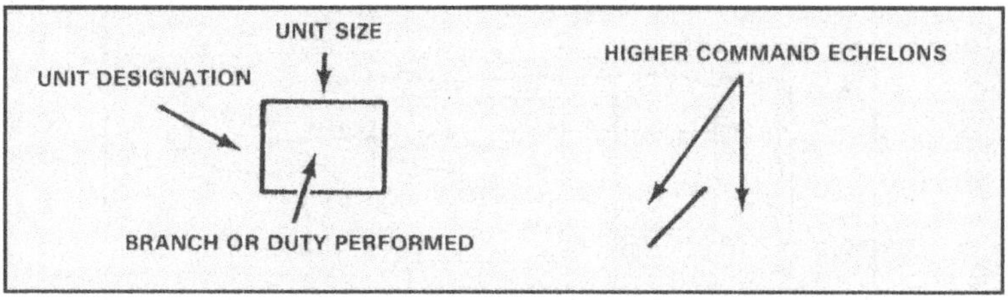


**Figure 071-329-1019-3. Development of unit symbol**

**c.** Unit-size symbols (figure 071-329-1019-4).

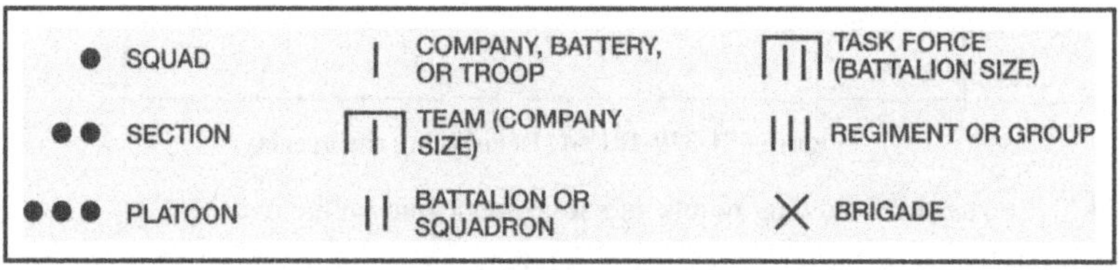


**Figure 071-329-1019-4. Unit-size symbols**

**d.** Branch symbols (figure 071-329-1019-5).

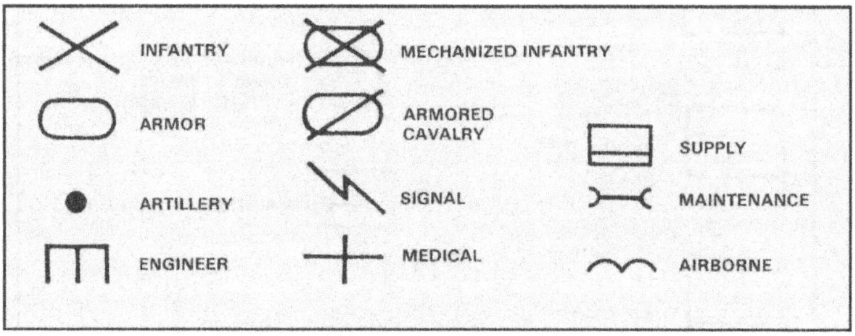


**Figure 071-329-1019-5. Branch symbols**

**e.** Enemy unit (red or double lines; figure 071-329-1019-6).

**Figure 071-329-1019-6. Enemy unit**

**Performance Steps**

f. Proposed or future unit position (broken lines; figure 071-329-1019-7).

**Figure 071-329-1019-7. Proposed unit position**

g. Tactical control symbols (figure 071-329-1019-8).

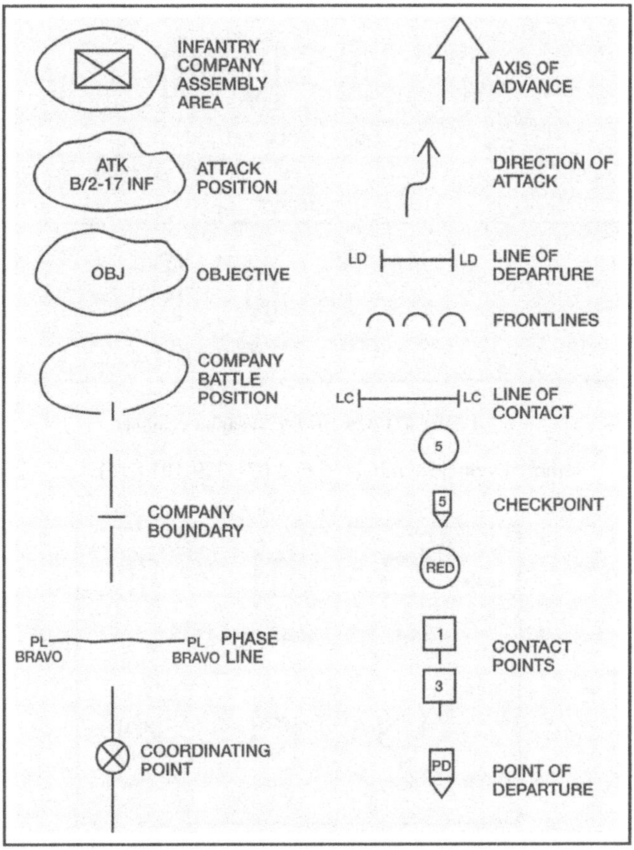


**Figure 071-329-1019-8. Tactical control symbols**

**Performance Steps**

**h.** Weapons symbols (figure 071-329-1019-9).

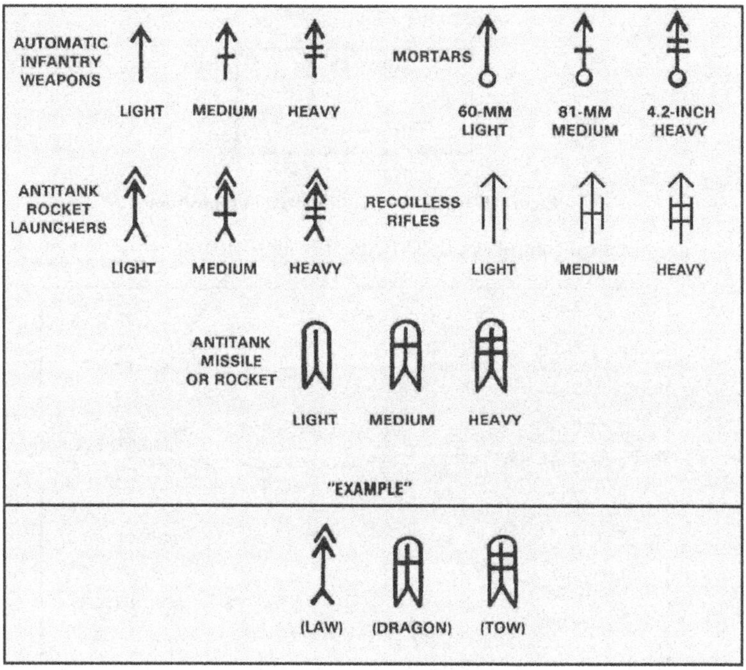


**Figure 071-329-1019-9. Weapon symbols**

**i.** Armored vehicle symbols (figure 071-329-1019-10).

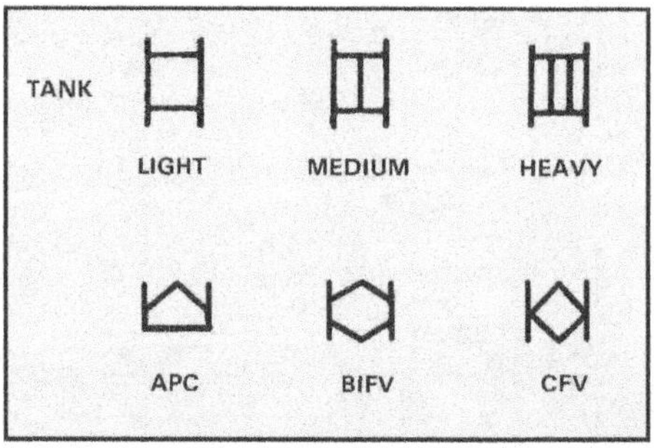


**Figure 071-329-1019-10. Armored vehicle symbols**

j. Fortification and obstacle symbols (figure 071-329-1019-11).

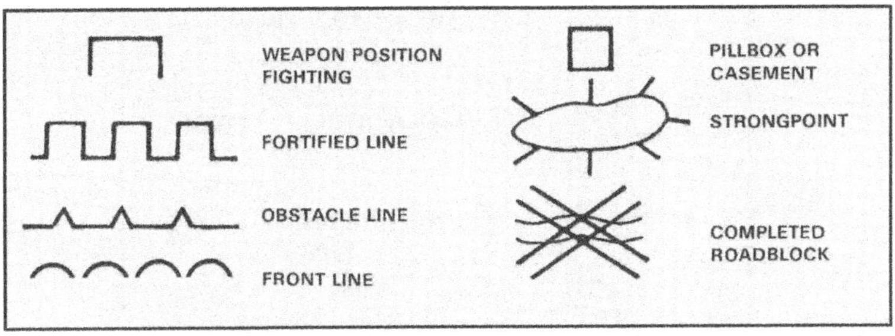


**Figure 071-329-1019-11. Fortification and obstacle symbols**

k. Tactical wire symbols (figure 071-329-1019-12).

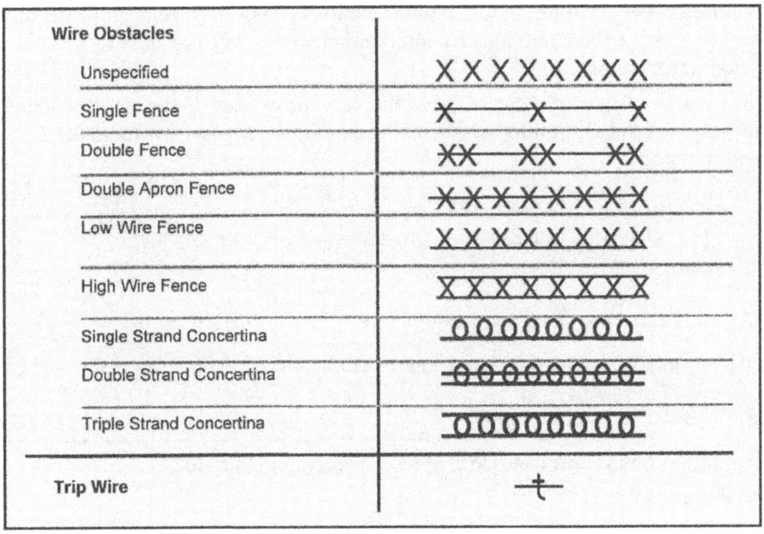


**Figure 071-329-1019-12. Tactical wire symbols**

l. Mine symbols (figure 071-329-1019-13).

**Figure 071-329-1019-13. Mine symbols**

| Performance Steps |
|---|

**m.** Indirect fire symbols (figure 071-329-1019-14).

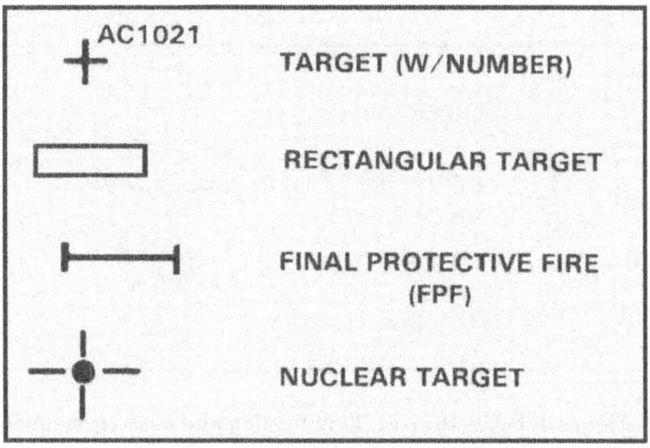


**Figure 071-329-1019-14. Indirect fire symbols**

**Evaluation Preparation:** *Setup*: Provide a military map and a company level map overlay. Ask the Soldier to identify information on the overlay from the performance measures.

*Brief Soldier:* Tell the Soldier to correctly place the overlay on the map and identify any graphic symbols or information on the overlay as requested, without the aid of any references.

| Performance Measures | GO | NO GO |
|---|---|---|
| **1.** Placed overlay grid register marks over grid intersections. | —— | —— |
| **2.** Identified marginal information. | —— | —— |
| **3.** Identified security classification. | —— | —— |
| **4.** Identified graphic symbols. | —— | —— |

**Evaluation Guidance:** Refer to chapter 1, paragraph 1-4 b (6).

**References**

**Required:** FM 1-02 and FM 3-25.26

**Related:**

## SUBJECT AREA 6: COMMUNICATE

# 158-100-4003

### Communicate Effectively at the Direct Leadership Level

**Conditions:** As a leader, you receive information which needs to be communicated using the FM 6-22 leadership competency of communicates and its supporting behaviors and actions.

**Standards:** Communicate to an individual Soldier or group of Soldiers ensuring that the intent of the message was received and a shared understanding was reached.

**Performance Steps**

**1.** Receve a task or message to communicate.

**2.** Listen actively.

**a.** Listen and watch attentively.

**b.** Use verbal and nonverbal means to reinforce with the speaker that you are paying attention.

**c.** Reflect on new information before expressing views.

**3.** Determine information sharing strategies.

**a.** Determine the necessary information to share with subordinates.

**b.** Determine the purpose of the information.

**4.** Ensure that communication is sensitive to cultural factors.

**a.** Maintain awareness of communications customs, expressions, actions, or behaviors.

**b.** Demonstrate respect for others.

**5.** Employ engaging communication techniques.

**a.** State goals.

**b.** Speak enthusiastically.

**c.** Use gestures that are appropriate but not distracting.

**6.** Present recommendations so that others understand the advantages.

**a.** Use logic and relevant facts in dialogue.

**b.** Keep the conversation on track.

**c.** Express well thought-out and well-organized ideas.

**7.** Convey thoughts and ideas to ensure shared understandings.

**a.** Express thoughts and ideas to an individual Soldier or a group of Soldiers.

**b.** Use appropriate means for communicating a message.

**c.** Communicate clearly and concisely.

**d.** Clarify when there is a question.

**Evaluation Preparation:** *Setup:* Prepare a message or task that requires the Soldier to respond accurately, according to task standards, to the following measures.

*Brief Soldier:* Tell the Soldier that he or she will be required to correctly respond to all of the measures to receive a GO on the task.

| Performance Measures | GO | NO GO |
|---|---|---|
| **1.** Received a task or message to communicate. | —— | —— |
| **2.** Listened actively. | —— | —— |
| **a.** Listened and watched attentively. | | |
| **b.** Used verbal and nonverbal means to reinforce with the speaker that you paid attention. | | |
| **3.** Determined information sharing strategies. | —— | —— |
| **a.** Determined the necessary information to share with subordinates. | | |
| **b.** Determined the purpose of the information. | | |
| **4.** Ensured that communication was sensitive to cultural factors. | —— | —— |
| **a.** Maintained awareness of communications, customs, expressions, actions, or behaviors. | | |
| **b.** Demonstrated respect for others. | | |
| **5.** Employed engaging communication techniques. | —— | —— |
| **a.** Stated goals. | | |
| **b.** Spoke enthusiastically. | | |
| **c.** Used gestures that were appropriate but not distracting. | | |
| **6.** Presented recommendations so others understood advantages. | —— | —— |
| **a.** Used logic and relevant facts in the dialogue. | | |
| **b.** Kept the conversation on track. | | |
| **c.** Expressed well thought-out and well-organized ideas. | | |
| **7.** Conveyed thoughts and ideas to ensure shared understandings. | —— | —— |
| **a.** Expressed thoughts and ideas to an individual Soldier or a group of Soldiers. | | |
| **b.** Used appropriate means for communicating a message. | | |
| **c.** Communicated clearly and concisely. | | |
| **d.** Clarified when there was a question. | | |

**Evaluation Guidance:** Refer to chapter 1, paragraph 1-4 b (6).

**References**

**Required:**

**Related:** FM 6-22 (FM 22-100)

---

## 158-100-4009

### Communicate in Writing

**Conditions:** Given a topic, mission requirements, and access to reference materials, AR 25-50, DA Pam 600-67, and FM 6-22

**Standards:** Write a memorandum, letter, or e-mail to inform, persuade, or direct others in order to aid in effective and efficient communication or decisionmaking.

| Performance Steps |
|---|

1. Define the writing requirement.

   a. State the purpose.

   b. Define the main point.

   c. Choose a format.

2. Gather all relevant information to support your topic or mission.

   a. Seek guidance, if required.

   b. Make an outline of ideas.

3. Write the draft.

   a. Organize ideas and information in a logical order.

   b. Determine the necessary information to include based on topic or mission and expected outcome.

   c. Incorporate the information/ideas into the selected format.

   d. Ensure that communication is sensitive to cultural factors.

4. Edit the draft applying the Army writing style.

   a. Review the standards for Army writing.

   b. Apply the general rules for constructing military correspondence.

   c. Apply active voice writing techniques.

5. Finalize the document.

   a. Determine if the document communicates the following:

      (1) States the goals for others to act on.
      (2) States the relevance to the audience and the purpose.
      (3) Clearly conveys the message.
      (4) Provides adequate information and facts.

   b. Make corrections if necessary.

6. Conduct a final review.

   a. Assess your document from the reader's point of view.

   b. Obtain and evaluate feedback from others on—

      (1) Coherence and organization of the information.
      (2) Paragraph and sentence organization. It must be clear and focus on the information.
      (3) Correct grammar and punctuation usage.

   c. Rewrite to clarify if there is a question.

7. Give the document to a supervisor for review and distribution.

---

**Evaluation Preparation:** *Setup:* Prepare a message or task that requires the Soldier to respond accurately, according to task standards, to the following performance measures.

*Brief Soldier:* Tell the Soldier that he or she will be required to correctly respond to all of the performance measures to receive a GO on the task.

| Performance Measures | GO | NO GO |
|---|---|---|
| 1. Defined the writing requirement. | —— | —— |
| a. Stated the purpose. | | |
| b. Defined the main point. | | |
| c. Chose a format. | | |
| 2. Gathered all relevant information to support your topic. | —— | —— |
| a. Sought guidance as required. | | |
| b. Made an outline of ideas. | | |
| 3. Wrote the draft applying the Army writing style. | —— | —— |
| a. Organized ideas and information in a logical order. | | |
| b. Determined the necessary information to include based on topic and outcome. | | |
| c. Reviewed the standards for Army writing. | | |

| Performance Measures | GO | NO GO |
|---|---|---|
| **d.** Applied the general rules for constructing military correspondence. | | |
| **e.** Applied active voice writing techniques. | | |
| **f.** Incorporated the information/ideas into the selected format. | | |
| **4.** Edited the draft applying the Army writing style. | —— | —— |
| **a.** Reviewed the standards for Army writing. | | |
| **b.** Applied the general rules for constructing military correspondence. | | |
| **c.** Applied active voice writing techniques. | | |
| **5.** Conducted the final review. | —— | —— |
| **a.** Assessed your document from the reader's point of view. | | |
| **b.** Obtained and evaluated feedback from others on— | | |
| **(1)** Coherence and organization of the information. | | |
| **(2)** Paragraph and sentence organization. It was clear and focused on the information. | | |
| **(3)** Correct grammar and punctuation usage. | | |
| **6.** Gave the document to a supervisor for review and distribution. | —— | —— |

**Evaluation Guidance:** Refer to chapter 1, paragraph 1-4 b (6).

**References**

**Required:**

**Related:** AR 25-50, DA Pam 600-67, FM 6-22, FM 6-22 (FM 22-100), and ST 22-2

**SUBJECT AREA 22: UNIT OPERATIONS**

## 071-326-5502

### Issue a Fragmentary Order

**Conditions:** Given changes to original operation order, and a requirement to develop and issue a fragmentary order (FRAGO).

**Standards:** Develop and issue a clear and brief FRAGO based on changes in the mission or additional information and issued the FRAGO in the standard operation order (OPORD) format.

**Performance Steps**

*Note*: The FRAGO provides timely changes of existing orders. A FRAGO is either oral or written and addresses only those parts of the original OPORD that have changed.

**1.** Use standard military terminology.

**2.** Issue the FRAGO in the same sequence as the OPORD. Use all five paragraph headings (figure 071-326-5502-1).

FRAGMENTARY ORDER ____________

References: (Mandatory) Reference the order being modified.

Time Zone Used Throughout the Order: (Optional)

1. SITUATION. (Mandatory) Include any changes to the existing order.

2. MISSION. (Mandatory) List the new mission.

3. EXECUTION.

   Intent: (Optional)

   a. Concept of operations. (Mandatory)

   b. Tasks to subordinate units. (Mandatory)

   c. Coordinating instructions. (Mandatory) Include the statement, "Current overlay remains in effect" or "See change 1 to Annex C, Operations Overlay." Mark changes to control measures on overlay or issue a new overlay.

4. SERVICE SUPPORT. Include any changes to existing order or the statement, "No changes to OPORD ___."

5. COMMAND AND SIGNAL. Include any changes to existing order or the statement, "No changes to OPORD ___."

ACKNOWLEDGE: (Mandatory)

NAME (Commander's last name)
RANK (Commander's rank)

**Figure 071-326-5502-1. FRAGO example**

**Evaluation Preparation:** *Setup*: At the test site, provide equipment as stated in the conditions statement.

*Brief Soldier:* Tell the Soldier to develop and issue a FRAGO that includes all changes to the original OPORD.

| Performance Measures | GO | NO GO |
|---|---|---|
| **1.** Issued a FRAGO. | ___ | ___ |
| **2.** Used standard military terminology. | ___ | ___ |
| **3.** Included all changes to the original OPORD. | ___ | ___ |

**Evaluation Guidance:** Refer to chapter 1, paragraph 1-4 b (6).

**References**

**Required:**

**Related:** FM 5-0

---

# 071-326-5503

**Issue a Warning Order**

**Conditions:** Given preliminary notice of an order or action that is to follow and a requirement to develop and issue a warning order (WARNORD) to subordinates.

**Standards:** Within time allotted, develop a warning order and issue it to subordinate leaders. Issue order so that all subordinate leaders understand their missions and any coordinated instructions. Issue it in the standard operation order (OPORD) format.

**Performance Steps**

**1.** Say "Warning order" before issuing the order.

**2.** Use standard terminology.

**3.** Issue the WARNORD in the five-paragraph OPORD format.

**a.** SITUATION paragraph.

**(1)** Enemy forces: Provide available information on disposition, composition, strength, capabilities, and most probable course of action.

**(2)** Friendly forces: Give available information about the missions of next higher and adjacent units.

**(3)** Attachments and detachments: Give information about any units that have been attached or detached.

**b.** MISSION paragraph. Clearly and concisely state the mission as a task to be accomplished and state the purpose for doing it.

**c.** EXECUTION paragraph. Provide information about the operation, if available.

**d.** SERVICE SUPPORT paragraph. Provide all known instructions and arrangements supporting the operation.

**e.** COMMAND AND SIGNAL paragraph. Designate the succession of command if it differs from the unit standing operating procedure (SOP).

***Note***: Warning orders involving movement should state the time of movement.

**Evaluation Preparation:** *Setup*: At the test site, provide a platoon-level warning order. The warning order will be given orally to the Soldier. The Soldier will then be given 10 minutes to prepare the squad warning order.

*Brief Soldier:* Tell the Soldier 1) to extract from the platoon warning order all information that pertains to the squad, 2) to prepare the squad warning order in 10 minutes, and 3) to present an oral squad warning order.

| Performance Measures | GO | NO GO |
|---|---|---|
| **1.** Said "Warning order." | —— | —— |
| **2.** Used standard military terminology. | —— | —— |
| **3.** Issued the warning order in the five-paragraph OPORD format. | —— | —— |
| **4.** Gave all available information. | —— | —— |

**Evaluation Guidance:** Refer to chapter 1, paragraph 1-4 b (6).

**References**

**Required:**

**Related:** FM 5-0

## 071-730-0006

### Enforce Operations Security

**Conditions:** In a combat environment, given a mission to conduct a tactical operation.

**Standards:** The enemy has been denied information on planned, ongoing, and completed operations. The unit practices camouflage, physical security, noise and light discipline, information security, authentication procedures, document security, sign and countersign, and terrain masking.

| Performance Steps |
|---|

**1.** Identify operations security requirements.

- **a.** Counter surveillance.
  - **(1)** Camouflage and concealment.
  - **(2)** Positions.
  - **(3)** Noise and light discipline.
- **b.** Physical security.
  - **(1)** Observation posts.
  - **(2)** Patrols.
  - **(3)** Stand to.
  - **(4)** Silent watch.
  - **(5)** Mounted and dismounted security.
- **c.** Signal security.
  - **(1)** Communications procedures.
  - **(2)** Electronic counter-countermeasures.
  - **(3)** Encoded and decoded information.

| Performance Steps |
|---|

**d.** Information security.

**(1)** Foreign nationals kept out of troop areas.

**(2)** Weapons and ammunition kept covered whenever possible.

**(3)** Vehicle markings and unit patches.

**(4)** Mail was censored.

**2.** Disseminate operational security information.

**3.** Make on-the-spot corrections.

**Evaluation Preparation:** *Setup:* At the test site, provide the leader with a mission to conduct a tactical operation.

*Brief Soldier:* Tell the Soldier to be familiar with camouflage, physical security, noise and light discipline, information security, authentication procedures, document security, sign and countersign, and terrain masking procedures.

| Performance Measures | GO | NO GO |
|---|---|---|
| **1.** Enforced counter surveillance. | —— | —— |
| **2.** Enforced physical security. | —— | —— |
| **3.** Enforced signal security. | —— | —— |
| **4.** Enforced information security. | —— | —— |
| **5.** Disseminated operational security information. | —— | —— |
| **6.** Made on-the-spot corrections as required. | —— | —— |

**Evaluation Guidance:** Refer to chapter 1, paragraph 1-4 b (6).

**References**

**Required:**

**Related:** FM 7-7

## SUBJECT AREA 23: SECURITY AND CONTROL

# 301-371-1200
## Process Captured Materiel

**Conditions:** This task can be performed in field and garrison locations under all conditions. Given captured enemy materiel consisting of a map, signal operation instructions, unidentified equipment, complete with reference manuals, captured enemy equipment tags, captured document tags, a sheet marked technical documentation (TECHDOC), blank spot reports, detainee's personal effects, and the unit standing operating procedure (SOP) describing how to process and evacuate captured enemy materiel (CEM).

**Standards:** Secure captured enemy materiel. Prepare a spot report on captured enemy materiel known or believed to be of intelligence value. Fill out and attach the tag to the item of captured enemy materiel. Receipt detainee's personal effects. Select correct disposition procedure for each item of captured enemy materiel.

**Performance Steps**

**1.** Safeguard the CEM.

**a.** Categorize the CEM into—

**(1)** Captured enemy documents (CEDs). This includes any piece of recorded information, regardless of form, obtained from the enemy, which subsequently comes into the hands of a friendly force. CEDs can be U.S. or allied documents that were once in the hands of the enemy.

**(2)** Captured enemy equipment (CEE). This includes all types of foreign materiel found on an enemy prisoner of war (EPW) or on the battlefield that may have military application.

**b.** Protect CEM.

**(1)** Protect CEM from looting, loss, misuse, recapture, or destruction by placing under guard at all stages during the intelligence exploitation process. Use the best method available to preserve the item in its original condition.

**(2)** Personal effects, to include money, will be receipted and a copy will be given to the detainee. Use DA Form 1132-R (*Prisoner's Personal Property List - Personal Deposit Fund*).

***Note:*** To ensure exploitation of the captured enemy materiel, all documenting paperwork and materiel must accompany the detainee to each holding facility. Follow EPW guidelines and the unit SOP.

***Note:*** Correctly identify documents to be impounded. Impound personal documents containing information of intelligence value that outweighs personal or sentimental value. Personal documents are those of a private or commercial origin, such as letters, diaries, newspapers, or books. Impounding means that the documents are taken with the intention of returning them to the EPW/detainee at a later time. Identification documents may be taken and briefly examined and then returned to the EPW/detainee. The Geneva Conventions (FM 27-10) state that identification documents may not be permanently removed from an EPW/detainee.

**2.** Prepare a spot report on all time-sensitive information.

**a.** Identify time sensitive information. The technical intelligence (TECHINT) appendix to an intelligence annex will delineate time-sensitive technical items and reporting channels.

***Note***: Time-sensitive information includes any significant intelligence information with either military or technical application, to include enemy order of battle, enemy maps, overlays, operation orders, signals, codes, cryptographic materiel, new weapons or equipment on the battlefield, information critical to successfully accomplishing friendly courses of action, or information that indicates a significant change in the enemy's capabilities or intentions.

**b.** Submit a spot report to notify higher commands of the capture or identification of enemy materiel believed to answer intelligence requirements.

**(1)** Use S-A-L-U-T-E format to identify the following:

**(a)** S–Size

**(b)** A–Activity

**(c)** L–Location

**(d)** U–Unit

**(e)** T–Time

**(f)** E–Equipment

**(2)** Report time-sensitive items by the least IMMEDIATE precedence.

**3.** Tag each CEM item.

**a.** Tag all CEM found on the battlefield with no known association to an EPW.

**(1)** Use CEE tag on each piece of equipment.

**(2)** Use a captured document tag for each document.

**b.** Attach a sheet marked TECHDOC to flag the CED, if the CED is associated with a particular piece of CEM.

**(1)** List the precise location, time, circumstances of capture, and a detailed description of the materiel on the sheet marked TECHDOC.

**(2)** Treat all documents marked TECHDOC with the highest priority and forward them through the intelligence officer to the higher command until their value is determined.

**c.** Use special procedures for handling captured communications-electronics (C-E) equipment and documents marked TECHDOC.

**(1)** Handle in close coordination with the communications staff.

**(2)** Treat CEDs containing communications or cryptographic information as secret material.

**(3)** Tag and evacuate through secure channels without delay.

**d.** Use the field expedient method when no preprinted tags are available. Include, as a minimum, the following information:

**(1)** Date and time of capture.

**(2)** Capturing unit and its designation.

**(3)** Circumstances of capture.

**(4)** Identity of the CED or CEE captured.

**(5)** Location where the document was captured, including the six- or eight-digit grid coordinate.

**e.** Tag without defacing the item.

**(1)** Ensure that all markings which appear on the materiel at the time of capture are preserved.

**(2)** Ensure that items are tagged so as not to damage or destroy the documents or equipment.

**(3)** Never write on the document itself.

**(4)** Put the tag, without damaging the CED, in a waterproof bag.

**(5)** Attach the tag to the CEM so that it will not come off.

**4.** Select the correct disposition procedure for each CEM item.

**a.** Identify all time-sensitive captured enemy materiel to be evacuated to the S2. The S2 will screen for intelligence or technical value and then evacuate to the rear for examination and exploitation. All enemy documents captured on the battlefield are sent immediately to the first intelligence staff officer in the chain. The S2/G-2 routes all enemy documents to the nearest interrogation element for tactical exploitation. Interrogators screen the documents for immediate information and forward them to higher command, as required.

**(1)** Evacuate CEM with any technical documents found that relate to its design or operation. When the materiel cannot be evacuated, the documents should be identified with the materiel by attaching a sheet marked TECHDOC and providing the following information:

**(a)** The precise location, time, circumstance of capture.

**(b)** A detailed description of the materiel.

**(c)** Photographs of the equipment (evacuate with the document).

**(d)** Photographs of the materiel, placing an object of known size such as a ruler near the materiel to provide a size reference.

***Note***: Protect equipment that is too large to evacuate. Detail guards to safeguard the item until you receive further instructions.

**(g)** Keep photos with TECHDOC and evacuate.

**(2)** Evacuate CEM through your chain of command to the S2. The S2 will screen for intelligence or technical value and then evacuate to the rear for examination and exploitation.

**b.** Evacuate other CEM through similar logistic assets (such as route captured petroleum, oil, and lubricants (POL) through your POL points).

**c.** Protect the CEM, which cannot be evacuated, until disposition instructions are received.

---

**Evaluation Preparation:** *Setup*: Prepare samples of captured enemy documents. Prepare captured document tags for CEDs. Prepare captured enemy equipment tags for all unidentified equipment, complete with reference manuals. Provide a sheet marked TECHDOC, blank spot reports, and the unit SOP.

*Brief Soldier*: Instruct the Soldier to process and evacuate all CEDs and CEE using captured document tags, captured enemy equipment tags, and a sheet marked TECHDOC when necessary.

| Performance Measures | GO | NO GO |
|---|---|---|
| **1.** Safeguarded each item. | —— | —— |
| **a.** Used the best method available to preserve the item in its original condition. | | |
| **b.** Protected the item from looting, loss, misuse, recapture, or destruction by placing it under guard at all stages during the intelligence exploitation process. | | |
| **c.** Receipted detainee's personal effects on DA Form 1132-R. | | |
| **2.** Submitted a written or verbal spot report on time-sensitive information that included six of six S-A-L-U-T-E items identified correctly. | —— | —— |
| **3.** Completed the appropriate tag for each item. | —— | —— |
| **a.** Used the captured enemy equipment or captured document tag for CEM found on the battlefield. | | |
| **b.** Attached a sheet marked TECHDOC to any CED associated with a particular piece of CEM. The cover sheet listed the precise location, time, circumstances of capture, and a detailed description of the materiel. Took photos of the CEM with a size reference, if possible, and kept the photo with the TECHDOC. | | |
| **c.** Use the field expedient method when no preprinted tags were available. | | |
| **d.** Tagged the item without defacing it. | | |
| **4.** Select the correct disposition procedure for each item. | —— | —— |

**Evaluation Guidance:** Refer to chapter 1, paragraph 1-4 b (6).

**References**

**Required:** DA Form 1132-R

**Related:** FM 2-22.3 (FM 34-52), FM 3-19.40 (FM 19-40), and FM 34-54

**SUBJECT AREA 25: EQUIPMENT CHECKS**

## 091-CTT-2001

**Supervise Preventive Maintenance Checks and Services (PMCS)**

**Conditions:** In a contemporary operational environment, given equipment, appropriate technical manuals, supporting forms and documentation, tools, petroleum, oils, and lubricants (POL) (if applicable), and personnel.

**Standards:** Ensure that the maintenance status of assigned equipment is reported and personnel are trained in the proper procedures for conducting preventive maintenance checks and services (PMCS) according to the applicable references.

**Performance Steps**

1. Direct PMCS.

**a.** Verify that all current and updated technical manuals and references are available or requisitioned for section's assigned equipment.

**b.** Verify that all tools, POL, personnel and other resources are available for PMCS.

**c.** Observe operators performing PMCS at prescribed intervals according to the technical manual.

**d.** Review maintenance forms and reporting procedures for accuracy and completeness.

**e.** Reduce training distractions.

2. Train personnel in the proper procedures for conducting PMCS.

**a.** Enforce the commander's training program for operators of assigned equipment.

**b.** Verify that all operators are licensed for their equipment.

**c.** Verify that all operators are properly trained with PMCS procedures.

3. Report the maintenance status of assigned equipment.

**a.** Verify that the operator has correctly identified and corrected or recorded faults on DA Form 5988-E (*Equipment Inspection Maintenance Worksheet*) or DA Form 2404 (*Equipment Inspection and Maintenance Work Sheet*).

**b.** Confirm that not mission capable (NMC) faults are corrected before dispatch (if applicable).

| Performance Measures | GO | NO GO |
|---|---|---|
| **1.** Verified that current and updated technical manuals and references were available. | ___ | ___ |
| **2.** Verified correctly that all tools, POL, personnel and other resources were available for PMCS. | ___ | ___ |
| **3.** Checked operator's DA Form 5988-E to confirm that operators were licensed. | ___ | ___ |
| **4.** Identified correct PMCS intervals according to the technical manual. | ___ | ___ |

| Performance Measures | GO | NO GO |
|---|---|---|
| **5.** Verified that the operator had correctly identified and corrected or recorded faults on DA Form 5988-E or DA Form 2404. | —— | —— |
| **6.** Reviewed maintenance forms and reporting procedures for accuracy and completeness. | —— | —— |

**Evaluation Guidance:** Refer to chapter 1, paragraph 1-4 b (6).

**References**

**Required:** AR 750-1, DA Pam 750-1, DA Pam 750-8, DA Pamphlet 750-35, and TM 9-243

**Related:**

## SUBJECT AREA 27: RISK MANAGEMENT

# 153-001-2000

## Employ the CRM Process and Principles and Show How They Apply to Performance of My Job/Assigned Duties

**Conditions:** As a small unit leader, given a mission, activity, or task, in a garrison or tactical environment, access to FM 5-19 and a DA Form 7566 (*Composite Risk Management Worksheet*).

**Standards:** Make an oral or written report that identifies the five steps of the composite risk management (CRM) process, and apply the CRM process to a given mission, activity, or task.

**Performance Steps**

**1.** Identify the five steps of the CRM process.

**2.** Identify and list hazards associated with the mission or task using mission, enemy, terrain and weather, troops and support available, time available, civil considerations (METT-TC) factors.

**3.** Assess the hazards to determine the level of risk (E,H,M,L). for each hazard and its impact on the mission or task

**4.** Select appropriate controls and reassess risk (E,H,M,L) for each identified hazard.

**5.** Identify appropriate methods for implementing the identified controls.

**6.** Identify appropriate personnel to supervise and evaluate the controls.

**7.** Determine the overall level of risk (E,H,M,L ) for the mission, activity, or task.

**Evaluation Preparation:** *Setup*: To evaluate this task, you need an assigned mission, task, or activity, on or off duty, which may be in the form of an operations order (OPORD), fragmentary order (FRAGO), warning order, patrol order, training task, and so forth

*Brief Soldier:* Tell the Soldier to identify the five steps of the CRM process and complete the CRM process for the assigned mission, task, or activity.

| Performance Measures | GO | NO GO |
|---|---|---|
| 1. Identified all five steps of the CRM process. | —— | —— |
| 2. Identified all hazards associated with the mission or task using METT-TC factors (a minimum of one, depending on the mission or task). | —— | —— |
| 3. Determined the level of risk (E,H,M,L) for each hazard. | —— | —— |
| 4. Selected appropriate controls for each hazard and reassessed the risk levels (E,H,M,L) for all identified hazards. | —— | —— |
| 5. Identified appropriate methods for implementing controls. | —— | —— |
| 6. Identified the appropriate personnel to supervise and evaluate control implementation. | —— | —— |
| 7. Determined the overall level of risk (E,H,M,L) for the mission, activity, or task. | —— | —— |

**Evaluation Guidance:** Refer to chapter 1, paragraph 1-4 b (6).

**References**

**Required:** FM 5-19, FM 5-0, and DA Form 7566

**Related:**

## SUBJECT AREA 28: ADMINISTRATION/MANAGEMENT

# 805C-PAD-2044

### Recommend Individual for Award

**Conditions:** You are a supervisor and have had the opportunity to observe and evaluate the performance of your subordinates. As a supervisor, you may have the opportunity to observe Soldiers performance or act of heroism, which warrant recognition. You have access to AR 600-8-22 and DA Form 638 (*Recommendation for Award)*, pen and/or computer with authorized software.

**Standards:** Submit the recommendation for award to appropriate approval authority (within established time lines.) Verify accuracy of Soldiers' personnel data and ensure justifications (achievements) are entered in bullet format.

| Performance Steps |
| --- |

**1.** Determine if the Soldier's performance or action merits an award.

**2.** Determine the type of award merited.

**a.** Review authorized awards.

**b.** Compare performance or act against award criterion.

**3.** Determine criteria for submitting award recommendations.

**a.** Identify the time limitations for submitting an award.

**b.** Identify the period of an award.

**c.** Identify the service, achievement, or heroism.

**d.** Verify nonduplication of awards.

**e.** Identify interim awards (if given).

**f.** Identify the rules for recognition upon retirement.

**g.** Identify the approving authorities.

**4.** Prepare the award recommendation (DA Form 638).

**a.** Complete the personal data on Soldier.

**b.** Specify the type and level of award.

**c.** Enter the period of service or date for which the award is being submitted.

**d.** Enter the supporting comments in block 20.

**e.** Attach the supporting documents required for awards for heroism.

**5.** Sign DA Form 638 in the appropriate block.

**6.** Submit the completed award recommendation to individual's chain of command.

**Evaluation Preparation:** *Setup:* To evaluate this task gather the items listed in the conditions statement. Provide the Soldier with sufficient information to prepare the personal data, type of award, and justification/achievement. (***Note:*** If your command directs use of PureEdge or other forms software, have the computer and printer available.)

*Brief Soldier:* Tell the Soldier that he/she will be evaluated on his/her ability to recommend an individual for an award.

| Performance Measures | GO | NO GO |
|---|---|---|
| **1.** Determined if the Soldier's performance or action merited an award. | —— | —— |
| **2.** Determined the type award merited. | —— | —— |
| **3.** Determined the criteria for submitting an award recommendation. | —— | —— |
| **a.** Identified the time limitations for submitting the award. | | |
| **b.** Identified the period of the award. | | |
| **c.** Identified the service, achievement, or heroism. | | |
| **d.** Identified if duplication of an award. | | |
| **e.** Identified interim award. | | |
| **f.** Identified the rules for recognition upon retirement. | | |
| **g.** Identified the recommendation official. | | |
| **h.** Identified the approving authorities. | | |
| **4.** Prepared the award recommendation. | —— | —— |
| **a.** Completed the personal data on the Soldier. | | |
| **b.** Specified the type and level of award. | | |
| **c.** Entered the period covered. | | |
| **d.** Entered the supporting comments in block 20. | | |
| **5.** Signed DA Form 638 in the appropriate block. | —— | —— |
| **6.** Forwarded DA Form 638 to the individual's chain of command. | —— | —— |

**Evaluation Guidance:** Refer to chapter 1, paragraph 1-4 b (6).

**References**

**Required:** AR 600-8-22 and DA Form 638

**Related:**

## 805C-PAD-2060

### Report Casualties

**Conditions:** You are a Soldier deployed in a field environment. Events are occurring that result in Soldier and Department of Defense (DOD) civilian casualties within U.S. and coalition forces. Given a casualty, DA Form 1156 (*Casualty Feeder Card*), a map, and a pen or pencil.

**Standards:** Legibly and accurately record all known data elements on the DA Form 1156. Submit the completed DA Form 1156 to your chain of command as soon as the tactical situation permits.

| Performance Steps |
|---|

**1.** Prepare DA Form 1156. Data items with asterisk, addressed below, indicate minimum information required to submit a casualty report. Complete as much of the DA Form 1156 as you can to preserve background information for the unit and family. (Data fields are numbered to align with the DCIPS-FWD.)

**(1)** Complete item 2 with the casualty type (such as hostile, nonhostile).

**(2)** Complete item 3 with the casualty's status (such as NSI, KIA, DUSTWUN).

***Note***: DUSTWUN/missing/captured will include date, time, and place last seen, age, height, weight, eyes, and identifying marks.

**(3)** Complete item 7 with the personnel type of casualty (such as military, civilian).

**(4)** Complete item 8 with the casualty's social security number (SSN) (9 digit number).

**(5)** Complete item 9 with the casualty's name (last, first, middle initial.).

**(6)** Complete item 13 with the service branch of the casualty.

**(7)** Complete item 14(a) with the name of the unit that the casualty is assigned to.

**(8)** Complete item 36 with the incident date/time.

**(9)** Complete item 37 with the place of the incident.

**(10)** Complete item 39 with the circumstances (detailed, factual account).

**(11)** Complete item 40 with the inflicting force and weapons.

**(12)** Enter other helpful information.

**(a)** Body armor and physical features.

**(b)** Identification of remains will include the means of identification: ID tags, name tag, personal recognition, other.

**2.** Have DA Form 1156 authenticated. Field grade commander (CDR) will authenticate the completed card for accuracy and thoroughness of items 2, 39, and 40 before dispatching the report. An authorized field grade officer may be designated to authenticate for the CDR.

**Evaluation Preparation:** *Setup*: Provide the Soldier with DA Form 1156, a map, pencil and paper. Give the Soldier a scenario that provides all the information needed to accomplish the performance measures.

*Brief Soldier*: Tell the Soldier that he/she will be evaluated on his/her ability to report casualties by preparing DA Form 1156.

| Performance Measures | GO | NO GO |
|---|---|---|
| **1.** Prepared DA Form 1156. | —— | —— |
| **a.** Completed item 2 with the casualty type. | | |
| **b.** Completed item 3 with the casualty's status. | | |

| Performance Measures | GO | NO GO |
|---|---|---|
| **c.** Completed item 7 with the personnel type of casualty. | | |
| **d.** Completed item 8 with the casualty's SSN. | | |
| **e.** Completed item 9 with the casualty's name. | | |
| **f.** Completed item 13 with the service branch of the casualty. | | |
| **g.** Completed item 14(a) with the name of the unit that the casualty is assigned to. | | |
| **h.** Completed item 36 with the incident date/time. | | |
| **i.** Completed item 37 with the place of the incident. | | |
| **j.** Completed item 39 with the circumstances. | | |
| **k.** Completed item 40 with the inflicting force and weapons. | | |
| **l.** Entered other helpful information. | | |
| **2.** Had a field grade CDR authenticate the completed card for accuracy and thoroughness of items 2, 39, and 40 prior to dispatch of report. | —— | —— |

**Evaluation Guidance:** Refer to chapter 1, paragraph 1-4 b (6).

**References**

**Required:** DA Form 1156 and FM 3-34.112

**Related:** AR 600-8-1

---

## 805C-PAD-2145

### Counsel a Soldier on the Contents of a Noncommissioned Officer Evaluation Report and NCOER Checklist

**Conditions:** You are a section leader, given noncommissioned officers (NCOs), mission statement or function manual, job descriptions, DA Form 2166-8-1 (*Noncommissioned Officer Counseling and Support Form*), DA Form 2166-8 (*Noncommissioned Officer Evaluation Report*), AR 623-3, DA Pam 623-3, FM 6-22, and FM 7-22.7.

**Standards:** Conduct initial performance counseling within 30 days of the NCO's assignment. Conduct quarterly performance counseling within specified times thereafter, or as performance warrants. Maintain counseling record on DA Form 2166-8-1.

**Performance Steps**

**1.** Determine the type of counseling, (for example, initial, quarterly, or as required). All counseling sessions are to be face-to-face encounters with two-way dialog.

**a.** Initial counseling is required within the first 30 days of the rating period.

**b.** Quarterly counseling is conducted after the initial counseling throughout the rating period.

**c.** Counseling sessions are also held as required, when the NCO's performance merits counseling outside of recurring time lines.

**2.** Gather support materials.

**a.** Gather the required forms.

**b.** Determine the duty descriptions.

**c.** Determine the mission statement.

**d.** Obtain the function manual.

**e.** Determine the areas of special emphasis.

**3.** Schedule the counseling.

**a.** Allot time for the Soldier to prepare.

**b.** Select the site.

**c.** Notify the individual.

**4.** Prepare for counseling.

**a.** Review the DA Form 2166-8-1 and related documents, such as other counseling records, requests for disciplinary action or recommendations for commendation.

**b.** Prepare the counseling outline/notes.

**5.** Conduct the counseling.

**a.** Provide individual assessment of his/her performance, if applicable.

**b.** Solicit input from the individual.

**c.** Provide the meaning of each value/responsibility on DA Form 2166-8.

**d.** Provide specific examples of excellence.

**e.** Determine the future counseling sessions.

**f.** Obtain the rated NCO's initials.

**Evaluation Preparation:** *Setup:* To evaluate this task, provide the Soldier a DA Form 2166-8, DA Form 2166-8-1, personal data sheet, job description (duties clearly stated), rating scheme, a scenario that specifies special areas of emphasis, AR 623-3, and DA Pam 623-3. Have the Soldier prepare the forms for an initial counseling to include personnel administrative data, listing of duty description, and special area of concern.

*Brief Soldier:* Tell the Soldier that he/she will be evaluated on his/her ability to conduct performance counseling on an NCOER. Tell the Soldier to prepare the forms for an initial counseling to include personnel administrative data, listing of duty description, and special area of concern.

| Performance Measures | GO | NO GO |
|---|---|---|
| **1.** Determined the type of counseling. | —— | —— |
| **2.** Gathered the support material. | —— | —— |
| **a.** Gathered the required forms. | | |
| **b.** Determined the duty description. | | |
| **c.** Determined the mission statement. | | |
| **d.** Obtained the function manual. | | |
| **e.** Determined the areas of special emphasis. | | |
| **3.** Scheduled counseling. | —— | —— |
| **a.** Allotted time for the Soldier to prepare. | | |
| **b.** Selected the site. | | |
| **c.** Notified the individual. | | |
| **4.** Prepared for the counseling. | —— | —— |
| **a.** Reviewed the DA Form 2166-8-1 and related documents. | | |
| **b.** Prepared the counseling outline/notes. | | |
| **5.** Conducted the counseling. | —— | —— |
| **a.** Provided individual assessment of his/her performance. | | |
| **b.** Solicited input from the individual. | | |
| **c.** Provided the meaning of each value/responsibility on DA Form 2166-8. | | |
| **d.** Provided the specific examples of excellence. | | |
| **e.** Determined future counseling sessions. | | |
| **f.** Obtained the rated NCO's initials. | | |

**Evaluation Guidance**: Refer to chapter 1, paragraph 1-4 b (6).

**References**

**Required:** AR 623-3, DA Form 2166-8, DA Form 2166-8-1, and DA Pam 623-3

**Related:** FM 6-22 and FM 7-22.7

# 805C-PAD-2407

## Recommend Disciplinary Action for a Soldier

**Conditions:** You are a squad/section leader. It has been observed or has been reported to you that one of your Soldiers has committed an act which may be in violation of standards, conduct, breach of discipline, or insubordination. Given access to a computer with appropriate software, printer, AR 25-50, AR 27-10, AR 600-8-2, AR 600-20, AR 601-280, AR 630-10 and AR 635-200.

**Standards:** Verify that the act/incident occurred. Determine if the conduct reported warrants disciplinary action. Submit the appropriate recommendation to your chain of command.

**Performane Steps**

**1.** Verify that the reported or observed act occurred.

**a.** Obtain written/sworn statements (if available).

**b.** Obtain blotter reports/counseling statements (if authorized).

**c.** Obtain the Soldier's perspective of the act/incident (if possible).

**2.** Determine if the act/incident is in violation of standards, professional conduct, breach of discipline, or insubordination.

**a.** Review all collected information.

**b.** Seek advice from the legal representative (unit legal clerk or JAG officer).

**c.** Make your decision based on the facts.

**3.** Prepare the recommendation in memorandum or as required by local directive.

**a.** Provide the summary of the act/incident.

**b.** Include all additional supporting documentation collected.

**c.** Provide a summary of the Soldier's current and past performance.

**d.** Provide your recommendation to the commander for his/her consideration and action.

**4.** Submit the recommendation to higher authority.

**Evaluation Preparation:** *Setup:* To evaluate this task, give the Soldier a scenario that would provide enough information to accomplish the performance steps. Provide access to a computer, printer, and references for the Soldier.

*Brief Soldier:* Tell the Soldier that he/she will be evaluated on his/her ability to recommend disciplinary action for a Soldier.

| Performance Measures | GO | NO GO |
|---|---|---|
| **1.** Verified that the reported or observed act occurred. | —— | —— |
| **a.** Obtained written/sworn statements. | | |
| **b.** Obtained blotter reports/counseling statements. | | |
| **c.** Obtained the Soldier's perspective of the act/incident. | | |
| **2.** Determined if the conduct was in violation of standards, breach of discipline, or insubordination. | —— | —— |
| **a.** Gathered all information | | |
| **b.** Sought advice from the legal representative. | | |
| **c.** Made the decision based on the facts. | | |
| **3.** Prepared and sent the recommendation to higher authority for disciplinary action. | —— | —— |
| **a.** Provided a summary of the act/incident. | | |
| **b.** Included all additional supporting documentation collected. | | |
| **c.** Provided a summary of the Soldier's current and past performance. | | |
| **d.** Provided a recommendation to the commander for his/her consideration and action. | | |
| **4.** Submitted recommendation to higher authority. | —— | —— |

**Evaluation Guidance:** Refer to chapter 1, paragraph 1-4 b (6).

**References**

**Required:** AR 25-50, AR 27-10, AR 600-20, AR 600-8-2, AR 601-280, AR 630-10, and AR 635-200

**Related:** FM 6-22

---

## 805C-PAD-2503

### Equal Opportunity Enforce Compliance with the Army's and Sexual Harassment Policies

**Conditions:** You are a small unit/section leader responsible for supervising personnel. Personnel include both male and female, and represent different races, colors, religions, and national origins. You have access to AR 600-20, FM 6-22 (FM 22-100), the Unit's and Army Equal Opportunity and Sexual Harassment Policies.

**Standards:** Demonstrate personal behavior and leadership consistent with the Army's equal opportunity (EO) and prevention of sexual harassment (POSH) policies. Enforce compliance with the Army's EO, POSH, and the Army's extremist activities policies.

**Performance Steps**

**1.** Act according to the Army's EO and prevention of sexual harassment policies.

**a.** Demonstrate Army values associated with EO.

**(1)** Display unquestionable loyalty.

**(2)** Follow your higher duty to the Army and the nation.

**(3)** Treat people as they should be treated.

**(4)** Live up to all the Army values.

**b.** Conform to the Army's EO and POSH policies. Do not—

**(1)** Make racial or sexual comments and/or gestures.

**(2)** Make national origin or religious comments/jokes/slurs.

**(3)** Display racist or sexually offensive visual materials.

**(4)** Make unsolicited and unwelcomed sexual contact with fellow Soldiers.

**(5)** Stereotype fellow Soldiers or make assumptions about their cultural background, race, religion, or beliefs.

**(6)** Use profanity or sexually oriented language.

**(7)** Discount the religious beliefs of fellow Soldiers.

**(8)** Belong to extremist organizations or participate in extremist activities.

**c.** Demonstrate leadership consistent with EO and POSH policies.

**(1)** Treat subordinates with dignity and respect.

**(2)** Treat subordinates fairly and equally.

**(3)** Recognize and respect subordinates' individual needs, aspirations, and capabilities.

**(4)** Do not discriminate against subordinates based on race, color, national origin, gender or religion.

**(5)** Do not use language that demeans, excludes, or offends subordinates.

**2.** Keep your unit free from unlawful discrimination and sexual harassment.

**a.** Ensure that subordinates understand Army and unit EO and POSH policies and procedures.

**(1)** Ensure that subordinates understand the required standards of behavior.

**(2)** Counsel subordinates on the legal and administrative repercussions of EO/sexual harassment violations.

**(3)** Ensure that subordinates understand recommended techniques for dealing with sexual harassment.

| Performance Steps |
|---|
| **(4)** Ensure that subordinates understand the EO and sexual harassment complaint process. |
| **(5)** Ensure that subordinates understand the Army policy on extremist activities and organizations. |
| **b.** Make on-the-spot corrections of subordinates whose behaviors are contrary to Army EO and sexual harassment policies. |
| **(1)** Correct Soldiers using racial or sexually harassing nonverbal gestures. |
| **(2)** Correct Soldiers using racial or sexually harassing verbal comments. |
| **(3)** Correct Soldiers displaying racial or sexually harassing visual materials. |
| **(4)** Do not allow intimidation, harassment, or reprisal against Soldiers for making complaints. |

**Evaluation Preparation:** *Setup*: This task can be tested in an administrative or field environment. Give the Soldier a scenario that provides sufficient information to evaluate his/her knowledge of the Army's EO and POSH polices, prohibited acts, and actions that should be taken if a violation occurs.

*Brief Soldier:* Inform Soldier that they will be tested on his/her knowledge of the Army's EO and POSH polices, prohibited acts, and actions that should be taken if a violation occurs.

| Performance Measures | GO | NO GO |
|---|---|---|
| **1.** Acted according to the Army's EO and POSH policies. | —— | —— |
| **a.** Demonstrated Army values associated with EO. | | |
| **(1)** Displayed unquestionable loyalty. | | |
| **(2)** Followed your higher duty to the Army and the nation. | | |
| **(3)** Treated people as they should be treated. | | |
| **(4)** Lived up to all the Army values. | | |
| **b.** Conformed to the Army's EO and POSH policies. | | |
| **(1)** Did not make racial or sexual comments and/or gestures. | | |
| **(2)** Did not make national origin or religious comments/jokes/slurs. | | |
| **(3)** Did not display racist or sexually offensive visual materials. | | |
| **(4)** Did not make unsolicited or unwelcomed sexual contact with fellow Soldiers. | | |
| **(5)** Did not stereotype fellow Soldiers or make assumptions about their cultural background, race, religion, or beliefs. | | |
| **(6)** Did not use profanity or sexually oriented language. | | |

| Performance Measures | GO | NO GO |
|---|---|---|
| **(7)** Did not discount the religious beliefs of fellow Soldiers. | | |
| **(8)** Did not belong to extremist organizations or participate in extremist activities. | | |
| **c.** Demonstrated leadership consistent with EO and POSH policies. | | |
| **(1)** Treated subordinates with dignity and respect. | | |
| **(2)** Treated subordinates fairly and equally. | | |
| **(3)** Recognized and respected subordinates' individual needs, aspirations, and capabilities. | | |
| **(4)** Did not discriminate against subordinates based on race, color, national origin, gender or religion. | | |
| **(5)** Did not use language that demeaned, excluded, or offended subordinates. | | |
| **2.** Kept your unit free from unlawful discrimination and sexual harassment. | ___ | ___ |
| **a.** Ensured that subordinates understood Army and unit EO/POSH policies and procedures. | | |
| **(1)** Ensured that subordinates understood the required standards of behavior. | | |
| **(2)** Counseled subordinates on the legal and administrative repercussions of EO/sexual harassment violations. | | |
| **(3)** Ensured that subordinates understood recommended techniques for dealing with sexual harassment. | | |
| **(4)** Ensured that subordinates understood the EO and sexual harassment complaint process. | | |
| **(5)** Ensured that subordinates understood the Army policy on extremist activities and organizations. | | |
| **b.** Made on-the-spot corrections of subordinates whose behaviors were contrary to Army EO and POSH policies. | | |
| **(1)** Corrected Soldiers that used racial or sexually harassing nonverbal gestures. | | |
| **(2)** Corrected Soldiers that used racial or sexually harassing verbal comments. | | |
| **(3)** Corrected Soldiers that displayed racial or sexually harassing visual materials. | | |
| **(4)** Did not allow intimidation, harassment, or reprisal against Soldiers that made complaints. | | |

**Evaluation Guidance:** Refer to chapter 1, paragraph 1-4 b (6).

**References**

**Required:**

**Related:** AR 600-13, AR 600-20, DA Pam 350-20, DODDIR 7050.6, FM 6-22 (FM 22-100) and Manual-MCM

## 158-100-3012

### Correlate a Leader's Role in Character Development with Values and Professional Obligations

**Conditions:** You are a leader in the U.S. Army. As a leader, you must know and understand how values and professional obligations guide your way of life as a member of the Profession of Arms.

**Standards:** Identify the relationship between the Oath of Office; Oath of Enlistment; Code of Conduct; Warrior Ethos; National, Army, and personal values; and professional obligations with a leader's role in character development according to FM 6-22 and FM 1.

| Performance Steps |
|---|

**1.** Discuss the relationship between character and beliefs, and character and ethics.

**2.** Identify how values shape the development of personal character and character of subordinates.

**3.** Identify how professional obligations shape the development of personal character and character of subordinates.

**4.** Identify the components of the process used by leaders to develop character in subordinates.

**Evaluation Preparation:** *Setup:* Provide Soldier with the references listed below. Prepare a scenario that requires the Soldier to respond accurately, according to task standards, to the following performance measures. This may be presented orally or in writing.

*Brief Soldier:* Tell the Soldier that he or she will be required to correctly respond to at least 75 percent of the performance measures to receive a GO on the task.

| Performance Measures | GO | NO GO |
|---|---|---|
| **1.** Discussed the relationship between character and beliefs, and character and ethics. | ___ | ___ |
| **a.** Defined character and identified the importance of character development in leaders. | | |
| **b.** Identified the relationship between character and beliefs. | | |
| **c.** Identified the relationship between character and ethics. | | |

| Performance Measures | GO | NO GO |
|---|---|---|
| **2.** Identified how values shaped the development of personal character and character of subordinates. | —— | —— |
| **a.** Identified how personal values shaped character development. | | |
| **b.** Identified how National values shaped character development. | | |
| **c.** Identified how Army values shaped character development. | | |
| **3.** Identified how professional obligations shaped the development of personal character and character of subordinates. | —— | —— |
| **a.** Identified how professional obligations imposed by the Oath of Enlistment shaped character development. | | |
| **b.** Identified how professional obligations imposed by the Oath of Office shaped character development. | | |
| **c.** Identified how professional obligations imposed by the Code of Conduct shaped character development. | | |
| **d.** Identified how professional obligations imposed by the Warrior Ethos shaped character development. | | |
| **4.** Identified the process used by leaders to develop character in subordinates. | —— | —— |
| **a.** Identified education. | | |
| **b.** Identified reinforcement. | | |
| **c.** Identified internalization. | | |

**Evaluation Guidance:** Refer to chapter 1, paragraph 1-4 b (6).

**References**

**Required:**

**Related:** CMH Pub 71-25, DD Form 4, DD Form 71, FM 1, FM 6-22 (FM 22-100), FM 7-22.7, GTA 21-03-009, GTA 21-03-010, and MISC Pub 870-3

---

## 158-100-7003

### Counsel a Subordinate

**Conditions:** You are assigned to a leadership position. You observe a behavior or pattern of behavior in a subordinate that requires leader intervention, or are given a requirement to conduct counseling. You have access to the appropriate counseling form (DA Form 4856 [*Developmental Counseling Form*] or DA Form 2166-8-1

[*Noncommissioned Officer Counseling and Support Form*]), FM 6-22, AR 635-200, and AR 135-178, if required.

**Standards:** Correctly identify the need for counseling and/or type of counseling required. Prepare for the counseling session by: 1) selecting a suitable place, 2) scheduling a time, 3) notifying the subordinate in advance, 4) organizing information, 5) outlining the counseling session component,; 6) planning a counseling trategy, and 6) establishing the right atmosphere. Conduct the counseling session to include: 1) opening the session, 2) discussing the issues, 3) developing a plan of action, 4) documenting the counseling session, and 5) closing the session. Conduct a follow-up of the counseling session to assess the plan of action and adjust it if necessary.

| **Performance Steps** |
|---|

**1.** Identify the need for counseling.

**2.** Identify the type of counseling required.

- **a.** Organizational or regulatory guidance.
- **b.** Reception and integration of new personnel.
- **c.** A subordinate crisis.
- **d.** Referral of a subordinate to an appropriate agency or organization.
- **e.** Not recommended for promotion.
- **f.** Adverse separation counseling.
- **g.** Performance counseling.
- **h.** Professional growth counseling.

**3.** Prepare for the counseling session to include—

- **a.** Selecting a suitable location.
- **b.** Scheduling a time.
- **c.** Notifying the subordinate of the time and place in advance.
- **d.** Organizing information.
- **e.** Outlining the counseling session components.
- **f.** Planning the counseling strategy.
- **g.** Establishing the right atmosphere.

**4.** Conduct the counseling session to include—

- **a.** Opening the session.
- **b.** Discussing the issues.
- **c.** Developing a plan of action.

| Performance Steps |
| --- |
| **d.** Documenting the session. |
| **e.** Closing the session. |
| **5.** Conduct a follow-up of the counseling session that— |
| **a.** Provides support for implementation of the counseling session. |
| **b.** Assesses the plan of action. |
| **c.** Adjusts the plan of action if required. |

**Evaluation Preparation:** *Setup:* Provide the Soldier with the references listed below. This task is best evaluated using a role playing scenario and an assistant to act as the subordinate. Prepare a scenario that requires the Soldier to perform accurately, according to task standards, to the following performance measures. The scenario may be presented verbally or in writing.

*Brief Soldier:* Tell the Soldier that he or she will be required to receive a GO on all performance measures to receive a GO for this task.

| Performance Measures | GO | NO GO |
| --- | --- | --- |
| **1.** Indentified the need for counseling. | —— | —— |
| **2.** Identified the type of counseling required. | —— | —— |
| **a.** Organizational or regulatory guidance. | | |
| **b.** Reception and integration of new personnel. | | |
| **c.** A subordinate crisis. | | |
| **d.** Referral of a subordinate to an appropriate agency or organization. | | |
| **e.** Not recommended for promotion. | | |
| **f.** Adverse separation counseling. | | |
| **g.** Performance counseling. | | |
| **h.** Professional growth counseling. | | |
| **3.** Prepared for the counseling session. | —— | —— |
| **a.** Selected a suitable location. | | |
| **b.** Scheduled a time. | | |
| **c.** Notified the subordinate of the time and place in advance. | | |
| **d.** Organized information. | | |
| **e.** Outlined the counseling session components. | | |

| Performance Measures | GO | NO GO |
|---|---|---|
| **f.** Planned the counseling strategy. | | |
| **g.** Established the right atmosphere. | | |
| **4.** Conducted the counseling session. | —— | —— |
| **a.** Opened the session. | | |
| **b.** Discussed the issues. | | |
| **c.** Developed a plan of action. | | |
| **d.** Documented the session. | | |
| **e.** Closed the session. | | |
| **5.** Conducted a follow-up of the counseling. | —— | —— |
| **a.** Provided support for implementation of the counseling session. | | |
| **b.** Assessed the plan of action. | | |
| **c.** Adjusted the plan of action if required. | | |

**Evaluation Guidance:** Refer to chapter 1, paragraph 1-4 b (6).

**References**

**Required:**

**Related:** AR 135-178, AR 635-200, and FM 6-22 (FM 22-100)

---

## 158-100-7012

### Develop Subordinates

**Conditions:** You are assigned to a leadership position and given the requirement to develop your subordinates as outlined in FM 6-22.

**Standards:** Demonstrate competency by: 1) correctly assessing the developmental needs of subordinates, 2) conducting professional growth counseling resulting in an individual development plan (IDP), 3) ensuring that your actions encourage and support your subordinates' ability to grow, and, 4) correctly conducting the performanc counseling.

| Performance Steps |
|---|
| **1.** Assess developmental needs of subordinates. |
| **a.** Observe subordinates' performance in the core leader competencies. |
| **b.** Record observations. |
| **c.** Determine if the performances meet, exceed, or fall below expected standards. |

| Performance Steps |
| --- |

**2.** Conduct professional growth counseling.

  **a.** Inform subordinates of your observations.

  **b.** Get feedback from subordinates.

**3.** Assist subordinates in designing an IDP.

  **a.** Identify actions to correct weaknesses.

  **b.** Identify actions to sustain strengths.

  **c.** Obtain subordinates' agreement to the plan.

**4.** Develop subordinates on the job

  **a.** Provide opportunities on the job.

  **b.** Assign tasks to provide practice in areas of subordinates' weaknesses.

  **c.** Provide challenging, mission-oriented training to improve practice.

**5.** Create a positive learning environment.

**6.** Share relevant personal experience with subordinates.

**7.** Provide counseling, coaching, or matching (as required).

**8.** Conduct periodic performance counseling (as required).

  **a.** Review the IDP to assess subordinates' progress.

  **b.** Modify the IDP if necessary.

**Evaluation Preparation:** *Setup:* Provide the Soldier with the references listed below. Prepare a scenario that requires the Soldier to perform accurately, according to task standards, to the following performance measures. The scenario may be presented verbally or in writing.

*Brief Soldier:* Tell the Soldier that he or she will be required to receive a GO on all performance measures to receive a GO for this task.

| Performance Measures | GO | NO GO |
| --- | --- | --- |
| **1.** Assessed developmental needs of subordinates. | —— | —— |
| **a.** Observed subordinates' performance. | | |
| **b.** Recorded observations. | | |
| **c.** Determined if the performances met, exceeded, or fell below expected standards. | | |
| **2.** Conducted professional growth counseling that— | —— | —— |
| **a.** Informed subordinates of observations. | | |
| **b.** Solicited comments from subordinates. | | |

| Performance Measures | GO | NO GO |
|---|---|---|
| **3.** Assisted subordinates in designing an IDP that— | ___ | ___ |
| **a.** Identified actions to correct weaknesses. | | |
| **b.** Identified actions to sustain strengths. | | |
| **c.** Obtained subordinates' agreement to the plan. | | |
| **4.** Provided opportunities for on-the-job development that— | ___ | ___ |
| **a.** Allowed opportunities for cross training. | | |
| **b.** Assigned tasks to provide practice in areas of subordinates' weaknesses. | | |
| **c.** Provided challenging, mission-oriented training to improve practice. | | |
| **5.** Performed actions to create a positive learning environment that— | ___ | ___ |
| **a.** Informed subordinates of existing individual self-development programs. | | |
| **b.** Provided training opportunities to improve subordinates self-awareness, confidence, and competence. | | |
| **c.** Applied effective assessment and training methods. | | |
| **6.** Shared relevant personal experiences with subordinates. | ___ | ___ |
| **7.** Provided counseling, coaching, or mentoring (as required). | ___ | ___ |
| **8.** Conducted periodic performance counseling (as required) that— | ___ | ___ |
| **a.** Reviewed the IDP to assess subordinates' progress. | | |
| **b.** Modified the IDP if necessary. | | |

**Evaluation Guidance:** Refer to chapter 1, paragraph 1-4 b (6).

**References**

**Required:**

**Related:** FM 6-22

## 158-100-7015

### Develop an Effective Team

**Conditions:** You have been assigned to a leadership position and are given the requirement to implement effective team building techniques as outlined in FM 6-22.

**Standards:** Implement an initial plan to apply the team building techniques outlined in FM 6-22. Perform all leader actions throughout the formation, enrichment, and sustainment stages of team building and develop an assessment plan to measure team effectiveness. Develop a revised plan, if necessary, to correct any deficiencies.

| Performance Steps |
|---|

**1.** Upon assignment to a leadership position, implement a plan to apply effective team building techniques.

**2.** Perform leader actions during the formation stage of team building.

**a.** General team building actions:

**(1)** Implement an effective reception and orientation plan.

**(2)** Create learning experiences.

**(3)** Communicate expectations.

**(4)** Listen to and care for subordinates.

**(5)** Reward positive contributions.

**(6)** Set an example.

**b.** Deployment team building actions:

**(1)** Talk with each Soldier.

**(2)** Reassure Soldiers by providing a calm presence.

**(3)** Communicate vital safety tips.

**(4)** Provide a stable situation.

**(5)** Establish a buddy system.

**(6)** Help Soldiers deal with immediate problems.

**3.** Perform leader actions during the enrichment stage of team building.

**a.** General team building actions:

**(1)** Demonstrate and encourage trust.

**(2)** Reinforce desired group norms.

**(3)** Establish clear lines of authority.

**(4)** Establish goals.

**(5)** Identify and grow leaders.

**(6)** Train as a team for the mission.

**(7)** Build pride through accomplishment.

**Performance Steps**

**b.** Deployment team building actions:

**(1)** Demonstrate competence.

**(2)** Prepare as a unit for operations.

**(3)** Know the Soldiers.

**(4)** Provide stable unit climate.

**(5)** Emphasize safety for improved readiness.

**4.** Perform leader actions during the sustainment stage of team building.

**a.** General team building actions:

**(1)** Demonstrate trust.

**(2)** Focus on teamwork, training, and maintaining.

**(3)** Respond to subordinate problems.

**(4)** Devise training that is more challenging.

**(5)** Build pride and spirit.

**b.** Deployment team building actions:

**(1)** Observe and enforce sleep discipline.

**(2)** Sustain safety awareness.

**(3)** Inform Soldiers.

**(4)** Know and deal with Soldiers' perceptions.

**(5)** Keep Soldiers productively busy.

**(6)** Use in-process reviews (IPRs) and after-action reviews (AARs).

**(7)** Act decisively in face of panic.

**5.** Assess team effectiveness and identify all deficiencies.

**6.** Develop a revised plan, if necessary, to correct any deficiencies.

---

**Evaluation Preparation:** *Setup:* Provide the Soldier with the references listed below. Prepare a scenario that requires the Soldier to perform accurately, according to task standards, to the following performance measures. This may be presented orally or in writing. The Soldier should be evaluated against the performance measures listed dependent on whether the team building event is for general team building purposes or in preparation for a deployment.

*Brief Soldier:* Tell the Soldier that he or she will be required to correctly respond to at least 75 percent of the performance measures to receive a GO on the task.

| Performance Measures | GO | NO GO |
|---|---|---|
| **1.** Implemented a plan that included all team building techniques outlined in FM 6-22. | ___ | ___ |
| **2.** Performed all leader actions during the formation stage of team building. | ___ | ___ |
| **a.** General team building actions: | | |
| **(1)** Implemented an effective reception and orientation plan. | | |
| **(2)** Created learning experiences. | | |
| **(3)** Communicated expectations. | | |
| **(4)** Listened to and cared for subordinates. | | |
| **(5)** Rewarded positive contributions | | |
| **(6)** Set example. | | |
| **b.** Deployment team building actions: | | |
| **(1)** Talked with each Soldier. | | |
| **(2)** Reassured Soldiers by providing a calm presence. | | |
| **(3)** Communicated vital safety tips. | | |
| **(4)** Provided a stable situation. | | |
| **(5)** Established a buddy system. | | |
| **(6)** Helped Soldiers deal with immediate problems. | | |
| **3.** Performed all leader actions during the enrichment stage of team building. | ___ | ___ |
| **a.** General team building actions: | | |
| **(1)** Demonstrated and encouraged trust. | | |
| **(2)** Reinforced desired group norms. | | |
| **(3)** Established clear lines of authority. | | |
| **(4)** Established goals. | | |
| **(5)** Identified and grew leaders. | | |
| **(6)** Trained as a team for the mission. | | |
| **(7)** Built pride through accomplishment. | | |

**b.** Deployment team building actions:

**(1)** Demonstrated competence.

**(2)** Prepared as a unit for operations.

**(3)** Knew the Soldiers.

**(4)** Provided stable unit climate.

**(5)** Emphasized safety for improved readiness.

**4.** Performed all leader actions during the sustainment stage of team building. ____ ____

**a.** General team building actions:

**(1)** Demonstrated trust.

**(2)** Focused on teamwork, training, and maintainin3.

**(3)** Responded to subordinate problems.

**(4)** Devised training that was more challenging.

**(5)** Built pride and spirit.

**b.** Deployment team building actions:

**(1)** Observed and enforced sleep discipline.

**(2)** Sustained safety awareness.

**(3)** Informed Soldiers.

**(4)** Knew and dealt with Soldiers' perceptions.

**(5)** Kept Soldiers productively busy.

**(6)** Used IPRs and AARs.

**(7)** Acted decisively in face of panic.

**5.** Assessed team effectiveness and identified all deficiencies. ____ ____

**6.** Developed a revised plan (if required) to correct all identified deficiencies. ____ ____

**Evaluation Guidance:** Refer to chapter 1, paragraph 1-4 b (6).

**References**

**Required:** FM 6-22

**Related:**

## 158-100-8006

### Solve Problems Using the Military Problem Solving Process

**Conditions:** Given available information about a problem and FM 5-0, FM 6-0, and FM 6-22.

**Standards:** Apply each step of the military problem solving process to solve a problem.

| Performance Steps |
|---|
| **1.** Identfy the problem. |
| **2.** Gather information. |
| **3.** Develop criteria. |
| **4.** Generate possible solutions. |
| **5.** Analyze possible solutions. |
| **6.** Compare possible solutions. |
| **7.** Make and implement the decision. |

**Evaluation Preparation:** *Setup:* Provide the Soldier with the references listed below. Prepare a scenario that requires the student to respond accurately, according to task standards, to the following performance measures. This may be presented orally or in writing.

*Brief Soldier:* Tell the Soldier that he or she will be required to correctly respond to at least 75 percent of the performance measures to receive a GO on the task.

| Performance Measures | GO | NO GO |
|---|---|---|
| **1.** Identified the problem. | —— | —— |
| **a.** Recognized and defined the problem. | | |
| **b.** Developed a problem statement. | | |
| **c.** Developed a plan to resolve the problem. | | |
| **2.** Gathered information. | —— | —— |
| **a.** Identified facts bearing on the problem. | | |
| **b.** Developed assumptions bearing on the problem. | | |
| **c.** Evaluated opinions bearing on the problem. | | |
| **3.** Developed criteria. | —— | —— |
| **a.** Defined the desired end state. | | |
| **b.** Developed screening criteria. | | |
| **c.** Developed evaluation criteria. | | |

| | | |
|---|---|---|
| **4.** Generated possible solutions. | ____ | ____ |
| **a.** Generated options applying brainstorming techniques. | | |
| **b.** Generated options applying intuition. | | |
| **c.** Develop solution statements. | | |
| **5.** Analyzed possible solutions. | ____ | ____ |
| **a.** Developed quantitative and/or qualitative measures for each solution. | | |
| **b.** Applied screening criteria to possible solutions. | | |
| **c.** Applied evaluation criteria to possible solutions. | | |
| **6.** Compared possible solutions. | ____ | ____ |
| **a.** Developed the comparison technique to be applied. | | |
| **b.** Compared and contrasted possible solutions. | | |
| **7.** Made and executed your decision. | ____ | ____ |
| **a.** Determined the preferred solution. | | |
| **b.** Developed the findings and recommended solution for decision. | | |
| **c.** Analyzed the solution for effectiveness. | | |

**Evaluation Guidance:** Refer to chapter 1, paragraph 1-4 b (6).

**References**

**Required:** FM 5-0

**Related:** FM 6-0 and FM 6-22

This page intentionally left blank.

# 081-831-1058

## Supervise Casualty Treatment and Evacuation

**Conditions:** You are a Soldier deployed to a unit in a forward area. There are casualties that must be treated and evacuated to receive medical aid. A military vehicle (ground vehicle or rotary-wing aircraft) may be available. You may have a litter and straps (or materials to improvise them) to secure the casualty and other Soldiers available to assist in treatment and evacuation.

**Standards:** Ensure that self-aid/buddy aid is provided to the casualties, as appropriate. Ensure that the casualties are transported to medical aid or to a pickup site using an appropriate carry or, if other Soldiers are available, by litter. Ensure that the litters are properly loaded onto a military vehicle (ground vehicle or rotary-wing aircraft) without dropping or causing further injury to the casualties.

**Performance Steps**

*Note:* Perform these procedures when medical and combat lifesaver personnel are not available. As soon as medical personnel are available, assist them, as necessary, in treating and evacuating the casualties.

**1.** Evaluate the casualties according to the tactical situation. (See STP 21-1-SMCT, task 081-831-1001.)

*Note:* Tactical combat casualty care (TCCC) can be divided into three phases.

- Care under fire—you are under hostile fire and are very limited as to the care you can provide.
- Tactical field care—you and the casualty are relatively safe and no longer under effective hostile fire, and you are free to provide casualty care to the best of your ability.
- Combat casualty evacuation care—care is rendered during casualty evacuation (CASEVAC).

**2.** Coordinate treatment (self-aid/buddy aid) of the casualties according to the tactical situation and available resources.

**3.** Request medical evacuation (MEDEVAC). (See task 081-831-0101.)

**a.** Make contact.

**b.** Determine whether casualties must be moved or will be picked up at the current location. If they must be moved, continue with step 4. If they will not be moved, continue to monitor communications and go to step 6.

**4.** Move a casualty, if necessary, using a four-man litter squad.

*Note:* If military vehicles and litter materials are not available, continue with step 5.

*Note:* Four-man litter squad bearers should be designated with a number from 1 to 4. The litter bearer designated as #1 is the leader of the squad.

**Performance Steps**

**a.** Prepare the litter.

**(1)** Open a standard litter.

**(2)** Lock the spreader bars at each end of the litter with your foot.

**b.** Prepare the casualty.

**(1)** Place the casualty onto the litter.

**(2)** Secure the casualty to the litter with litter straps.

**c.** Lift the litter.

**(1)** Position one squad member at each litter handle with the litter squad leader at the casualty's right shoulder.

*Note:* The leader should be at the right shoulder to monitor the casualty's condition.

**(2)** On the preparatory command, "Prepare to lift," the four bearers kneel beside the litter and grasp the handles.

**(3)** On the command, "Lift," all bearers rise together.

**(4)** On the command, "Four man carry, move," all bearers walk forward in unison.

**(a)** If the casualty does not have a fractured leg, carry the casualty—

- Feet first on level ground.
- Head first when going up hill.

**(b)** If the casualty has a fractured leg, carry the casualty—

- Head first on level ground.
- Feet first when going up hill.

**(5)** To change direction of movement, such as from feet first to head first, begin in a litter post carry position. The front and back bearers release the litter and the middle bearers rotate the litter and themselves.

**5.** Coordinate transportation of casualties using appropriate carries, if necessary. (See STP 21-1-SMCT, task 081-831-1046.)

**6.** Coordinate loading of casualties onto a military vehicle.

**a.** Ground ambulance.

*Note:* Ground ambulances have medical personnel to take care of the casualties during evacuation. Follow any special instructions for loading, securing, or unloading casualties.

**(1)** Secure each litter casualty to his/her litter.

**(2)** Load the most serious casualty last.

**(3)** Load the casualty head first (head in the direction of travel) rather than feet first.

**(4)** Secure each litter to the vehicle.

**Performance Steps**

**b.** Air ambulance.

*Note:* Air ambulances have medical personnel to take care of the casualties during evacuation. Follow any special instructions for loading, securing, or unloading casualties.

**(1)** Remain 50 yards from the helicopter until the litter squad is signaled to approach the aircraft.

**(2)** Approach the aircraft in full view of the aircraft crew, maintaining visual confirmation that the crew is aware of the approach of the litter party. Ensure that the aircrew can continue to visually distinguish friendly from enemy personnel at all times. Maintain a low silhouette when approaching the aircraft.

**(3)** Approach UH-60/UH-1 aircraft from the sides. Do not approach from the front or rear. If you must move to the opposite side of the aircraft, approach from the side to the skin of the aircraft. Then, hug the skin of the aircraft, and move around the front of the aircraft to the other side.

**(4)** Load the most seriously injured casualty last.

**(5)** Load the casualty who will occupy the upper birth first, then load the next litter casualty immediately under the first casualty.

*Note:* This is done to keep the casualty from accidentally falling on another casualty if his/her litter is dropped before it is secured.

**(6)** When casualties are placed lengthwise, position them with their heads toward the direction of travel.

**(7)** Secure each litter casualty to his/her litter.

**(8)** Secure each litter to the aircraft.

**c.** Ground military vehicles.

*Note:* Nonmedical military vehicles may be used to evacuate casualties when no medical evacuation vehicles are available. If medical personnel are present, follow their instructions for loading, securing, or unloading casualties.

*Note:* FM 8-10-6 contains suggested loading plans for many common nonmedical vehicles. You should become familiar with the plans for vehicles assigned to your unit.

**(1)** When loading casualties into the vehicle, load the most seriously injured casualty last.

**(2)** When a casualty is placed lengthwise, load the casualty with his/her head pointing forward, toward the direction of travel.

**(3)** Secure each litter casualty to his/her litter.

**(4)** Secure each litter to the vehicle as it is loaded into place.

**(5)** Watch the casualties closely for life-threatening conditions and provide first aid, as necessary, during CASEVAC.

*Note:* CASEVAC refers to the movement of casualties aboard nonmedical vehicles or aircraft. Care is rendered while the casualty is awaiting pickup or is being transported. A Soldier accompanying an unconscious casualty should monitor the casualty's airway, breathing, and bleeding.

**Evaluation Preparation:** *Setup:* Evaluate this task during a training exercise involving a MEDEVAC aircraft or vehicle, or simulate it by creating a scenario, and provide the equipment needed for the evaluation.

*Brief Soldiers:* Tell the Soldiers the scenario to include the end result desired.

| Performance Measures | GO | NO GO |
|---|---|---|
| **1.** Evaluated the casualties according to the tactical situation. | ___ | ___ |
| **2.** Coordinated treatment (self aid/buddy aid) of the casualties according to the tactical situation and available resources. | ___ | ___ |
| **3.** Requested medical evacuation. | ___ | ___ |
| **4.** Moved a casualty using a four-man litter squad, if necessary. | ___ | ___ |
| **a.** Prepared the litter. | | |
| **b.** Prepared the casualty. | | |
| **c.** Lifted the litter. | | |
| **5.** Coordinated transportation of casualties using appropriate carries, if necessary. | ___ | ___ |
| **6.** Coordinated loading of casualties onto a military vehicle. | ___ | ___ |

**Evaluation Guidance:** Refer to chapter 1, paragraph 1-4 b (6).

**References**

**Required:**

**Related:** FM 4-02.2 (FM 8-10-26), FM 4-25.11 (FM 21-11), FM 8-10-6, and STP 21-1-SMCT

---

## 081-831-1059

### Implement Measures to Reduce Combat Stress

**Conditions:** You are a leader of a group of Soldiers preparing to enter or already in a combat situation.

**Standards:** Recognize stress-related behaviors and implement appropriate leader actions that offset and control combat and operational stress reaction risk factors.

| Performance Steps |
|---|
| **1.** Recognize stress-related behaviors. (See figure 081-831-1059-1.) |

**a.** Positive combat stress behaviors. Positive combat stress behaviors include the heightened alertness, strength, endurance, and tolerance to discomfort which the fight or flight stress response and the stage of resistance can produce when properly in tune.

**Performance Steps**

**b.** Misconduct stress behaviors. These range from minor breaches of unit orders or regulations to serious violations of the Uniform Code of Military Justice (UCMJ) and perhaps the Law of Land Warfare.

**c.** Combat and operational stress reaction (COSR) (previously called battle fatigue). Some COSR behaviors may accompany excellent combat performance and are often found in heroes. More serious behaviors are warning signs and deserve immediate attention by the leader, medic, or buddy to prevent potential harm to the Soldier, others, or the mission.

**d.** Post-traumatic stress disorder (PTSD). PTSD is a psychiatric disorder that can occur following the experience or witnessing of life-threatening events such as military combat, natural disasters, and terrorist incidents. Some people have stress reactions that do not go away on their own, or may even get worse over time. People who suffer from PTSD often relive the experience through nightmares and flashbacks, have difficulty sleeping, and feel detached or estranged. These symptoms can be severe enough and last long enough to significantly impair the Soldier's daily life.

Performance Steps

| COMBAT STRESS BEHAVIORS | | |
|---|---|---|
| ***Positive Combat Stress Behaviors*** | ***Misconduct Stress Behaviors and Criminal Acts*** | ***Combat and Operational Stress Reaction*** |
| Unit Cohesion<br>Loyalty to Buddies<br>Loyalty to Leaders<br>Identification with Unit Tradition<br>Sense of Eliteness<br>Sense of Mission<br>Alertness, Vigilance<br>Exceptional Strength and Endurance<br>Increased Tolerance to Hardship, Discomfort, Pain, and Injury<br>Sense of Purpose<br>Increased Faith<br>Heroic Acts<br>Courage<br>Self-Sacrifice | Mutilating Enemy Dead<br>Not Taking Prisoners<br>Killing Enemy Prisoners<br>Killing Noncombatants<br>Torture, Brutality<br>Killing Animals<br>Fighting with Allies<br>Alcohol and Drug Abuse<br>Recklessness, Indiscipline<br>Looting, Pillage, Rape<br>Fraternization<br>Excessively on Sick Call<br>Negligent Disease, Injury<br>Shirking, Malingering<br>Combat Refusal<br>Self-Inflicted Wounds<br>Threatening/Killing Own Leaders ("Fragging")<br>Going Absent Without Leave, Desertion | Hyperalertness<br>Fear, Anxiety<br>Irritability, Anger, Rage<br>Grief, Self-Doubt, Guilt<br>Physical Stress Complaints<br>Inattention, Carelessness<br>Loss of Confidence<br>Loss of Hope and Faith<br>Depression, Insomnia<br>Impaired Duty Performance<br>Erratic Actions, Outbursts<br>Freezing, Immobility<br>Terror, Panic Running<br>Total Exhaustion, Apathy<br>Loss of Skills and Memories<br>Impaired Speech or Muteness<br>Impaired Vision, Touch, and Hearing<br>Weakness and Paralysis<br>Hallucinations, Delusions |
| | ***Post-Traumatic Stress Disorder*** | |
| | Intrusive Painful Memories, "Flashbacks"<br>Trouble Sleeping, Bad Dreams<br>Guilt About Things Done or Not Done<br>Social Isolation, Withdrawal, Alienation<br>Jumpiness, Startle Responses, Anxiety<br>Alcohol or Drug Misuse, Misconduct | |

**Figure 081-831-1059-1. Combat stress behaviors**

**2.** Identify COSR risk factors.

**a.** Domestic worries.

**b.** Newly assigned to unit.

**c.** First time in combat, horrors.

**d.** Casualties.

**e.** Lack of mobility.

**f.** Surprise attacks.

**g.** Inability to strike back—indirect fire, improvised explosive devices (IEDs).

**Performance Steps**

**h.** Information vacuum.

**i.** Chemical, biological, radiological, and nuclear (CBRN) weapons threat.

**j.** Sleep loss.

**k.** Physical exhaustion.

**l.** Dehydration, hunger.

**3.** Take leader actions to prevent/control COSR.

**a.** Integrate unit members; build unit cohesion and pride.

**b.** Help Soldiers stabilize the home front.

**c.** Keep Soldiers physically fit.

**d.** Conduct tough, realistic training.

**e.** Cross train in key areas.

**f.** Enforce sleep discipline.

**g.** Plan for personal hygiene.

**h.** Preserve Soldiers' welfare, safety, and health.

**i.** Reduce uncertainty.

**j.** Enforce individual health protection measures.

**k.** Utilize battlemind to build resiliency and dispel stigma.

**4.** Take leader actions to manage COSR.

***Note***: When a Soldier requires medical attention to rule out a possible serious physical cause for his/her symptoms, or because his/her inability to function endangers himself/herself, the unit, and the mission, he/she should be transported to the battalion aid station (BAS) or equivalent nearest medical support facility.

**a.** If a Soldier's behavior endangers the mission, himself/herself, or others, take appropriate measures to control him/her.

**b.** If a Soldier's is upset, let him/her talk about what is upsetting him/her, listen, and then try to reassure him/her.

***Note***: The most effective treatment for COSR is to normalize the symptoms presented by the Soldier. A Soldier does not know how he/she will react to combat. An effective leader will ensure that a Soldier understands that there are many normal physical and emotional reactions. It is imperative that the small group leader also verbally and nonverbally illustrate that the expectation is for the Soldier to improve and rejoin his/her organization as a fully functioning member.

| Performance Steps |
|---|

**c.** If a Soldier's reliability becomes questionable—

**(1)** Unload the Soldier's weapon.

**(2)** Remove the weapon only if the Soldier's behavior endangers the mission, himself/herself, or others.

**(3)** Physically restrain the Soldier only when safety is a concern or during transport.

**(4)** Reassure unit members that the signs are probably a normal COSR reaction and will quickly improve.

**d.** If the combat and operational stress reaction signs continue—

**(1)** Get the Soldier to a safer place.

**(2)** Do not leave the Soldier alone. Keep someone he/she knows with him/her.

**(3)** Notify the senior noncommissioned officer (NCO) or officer.

**(4)** Have the Soldier examined by medical personnel.

***Note***: When COSR casualties cannot be managed in place, they should be moved to a safer, quieter place, and provided rest and work for several hours up to one to two days in a place controlled by the unit. If the unit cannot wait for the Soldier to recover, he/she must be moved to the first level medical supporting unit. From there, every effort is made to move the Soldier to a nonmedical unit or area (a tent or building of opportunity could suffice) for rest, replenishment, and reassurance.

**e.** If the tactical situation permits, give the Soldier simple tasks to do when not sleeping, eating, or resting.

**f.** Assure the Soldier that he/she will return to full duty as soon as possible.

---

**Evaluation Preparation:** *Setup*: Prepare a scenario that requires the Soldier to respond to questions about the performance measures. This may be presented orally or in writing.

*Brief Soldier:* Tell the Soldier that he/she will be required to correctly respond to questions about combat stress behaviors.

| Performance Measures | GO | NO GO |
|---|---|---|
| **1.** Recognized stress-related behaviors. | —— | —— |
| **a.** Positive combat stress behaviors. | | |
| **b.** Misconduct stress behaviors. | | |
| **c.** COSR. | | |
| **d.** Post-traumatic stress disorder. | | |

| Performance Measures | GO | NO GO |
|---|---|---|
| **2.** Accurately identified COSR risk factors. | —— | —— |
| **3.** Took appropriate leader actions to prevent/control COSR. | —— | —— |
| **4.** Took appropriate leader actions to manage COSR. | —— | —— |

**Evaluation Guidance:** Refer to chapter 1, paragraph 1-4 b (6).

**References**

**Required:**

**Related:** FM 22-51, TG 240, TG 241, and TG 242

## SUBJECT AREA 3: CHEMICAL, BIOLOGICAL, RADIOLOGICAL, AND NUCLEAR

# 031-503-1016

## Implement Mission-Oriented Protective Posture (MOPP)

**Conditions:** You are in a chemical, biological, radiological, and nuclear (CBRN) environment, or you are warned of a threat of an CBRN hazard. You are given two or more Soldiers with MOPP gear, M8 and M9 detector paper, an M291 or M295 decontaminating kit, three nerve-agent antidote autoinjectors, FM 3-11.4, and requirements to assume the appropriate MOPP level and check Soldiers sleeping in MOPP4.

**Standards:** Implement and direct Soldiers to assume the appropriate MOPP level, and identify and rectify all deficiencies for Soldiers sleeping in MOPP4.

| Performance Steps |
|---|

**1.** Direct Soldiers to put on gear and equipment for MOPP0 through MOPP4.

**a.** Ensure that Soldiers assume MOPP0.

**b.** Ensure that Soldiers assume MOPP1.

**c.** Ensure that Soldiers assume MOPP2.

**d.** Ensure that Soldiers assume MOPP3.

**e.** Ensure that Soldiers assume MOPP4.

**2.** Ensure that Soldiers are wearing the appropriate clothing and equipment properly for the directed MOPP level and that Soldiers don and seal protective masks at MOPP3 and MOPP4.

**3.** Check Soldiers sleeping in MOPP4.

**a.** Observe the Soldiers to ensure that they are breathing.

**b.** Try to wake the Soldiers if they do not appear to be breathing.

**c.** Perform the task Evaluate a Casualty (task 081-831-1000).

| Performance Steps |
|---|

4. Check the Soldiers' mask for indications of a broken seal.

**a.** Wake the Soldiers and have them reseal their masks if the seal appears to be broken.

**b.** Direct a Soldier to observe another Soldier for symptoms of nerve-agent poisoning.

**c.** Perform first aid for nerve-agent injuries if symptoms are observed.

5. Check the Soldiers' protective clothing for problems.

**a.** Fasten or adjust the protective clothing, if the Soldiers' skin is not exposed and no symptoms are observed.

**b.** Wake the Soldiers and have them decontaminate using the M291 if their skin is exposed or they appear to be contaminated with liquid. Have the Soldiers adjust their protective clothing or conduct a MOPP gear exchange, as appropriate.

**c.** Perform first aid for nerve-agent injuries if skin is exposed or nerve-agent symptoms are observed.

**Evaluation Preparation:** *Setup*: Evaluate this task during a normal training session. Gather all necessary MOPP gear, and ensure that it is in good condition. Be prepared to direct a series of specific MOPP levels for the evaluated Soldier to implement with the troops provided (such as MOPP0 through MOPP4, sequentially). Have the Soldier explain the deficiencies that he/she should look for and the corrective actions he/she should take for a Soldier sleeping in MOPP4.

*Brief Soldier:* Tell the Soldier that the test will consist of directing Soldiers to assume the appropriate MOPP level and identifying all deficiencies and taking the appropriate corrective actions for a Soldier sleeping in MOPP4.

| Performance Measures | GO | NO GO |
|---|---|---|
| **1.** Directed Soldiers to put on gear and equipment for MOPP0 through MOPP4. | — | — |
| **2.** Ensured that Soldiers were wearing the appropriate clothing and equipment properly for the directed MOPP level and that Soldiers donned and sealed protective masks at MOPP3 and MOPP4. | — | — |
| **3.** Checked Soldiers sleeping in MOPP4. | — | — |
| **4.** Checked the Soldiers' masks for indicators of a broken seal. | — | — |
| **5.** Checked the Soldiers' protective clothing for problems. | — | — |

**Evaluation Guidance:** Refer to chapter 1, paragraph 1-4 b (6).

**References**

**Required:** FM 3-11.4

**Related:** FM 4-25.11, TM 10-8415-209-10, and TM 3-4240-279-10

## 031-503-3004

### Supervise the Crossing of a Contaminated Area

**Conditions:** You receive orders to cross a chemical, biological, radiological, or nuclear (CBRN) contaminated area. You are given a unit with mission-oriented protective posture (MOPP) gear; organic decontamination equipment; individual decontaminating kits; an M256A1chemical-agent detector kit; M8 chemical-agent detector paper; M9 chemical-detector tape; AN/VDR-2 and AN/UDR-13 radiac sets; detection and/or warning devices; shielding material (such as sandbags); FM 3-11.4; FM 3-11.3; FM 3-11.5 (FM 3-5); and a defined CBRN contaminated area.

**Standards:** Supervise a unit crossing or passing through a CBRN contaminated area without producing additional casualties or spreading contamination.

**Performance Steps**

**1.** Supervise a unit crossing a nuclear contaminated area.

**a.** Before crossing—

**(1)** Provide shielding for personnel. Use vehicles if possible. Place sandbags on the floor and sides of all vehicles (within reason for nonarmored vehicles).

**(2)** Tell vehicle operators to close all doors, windows, hatches, and vents on vehicles.

**(3)** Have Soldiers cover all exposed skin by rolling down their sleeves and buttoning their collars. Ensure that they wear handkerchiefs or similar cloths over their noses and mouths to keep from breathing radioactive dirt or dust.

**(4)** Select the shortest possible route that would cause the least contamination and permit the fastest travel based on mission, enemy, terrain and weather, troops and support available, time available, civil considerations (METT-TC).

**(5)** Ensure that AN/UDR-13 and AN/VDR-2 radiac sets are available and operational.

**(6)** Delay entry into the area as long as possible, within the limits of the mission.

**b.** During crossing—

**(1)** Perform continuous monitoring.

**(2)** Have the monitors watch the dose rate on the AN/UDR-13 or AN/VDR-2 radiac set. Instruct them to keep you informed. Ensure that the commander's turn-back dose rate is not exceeded without approval.

**(3)** Have all personnel with dosimeters check them often to ensure that the commander's turn-back dose is not exceeded without approval.

**(4)** Move through the area as quickly as possible. Do NOT forget safety or security.

**Performance Steps**

**(5)** Ensure that vehicles are far enough apart during movement to minimize dust. Consider the tactical situation, command, and control when spacing the vehicles.

**c.** After crossing—

**(1)** Have personnel check themselves and their equipment for contamination. Have everyone brush the dust from his/her clothing.

**(2)** Determine if decontamination is required.

**2.** Supervise a unit crossing a chemical contaminated area.

**a.** Before crossing—

**(1)** Select the shortest possible route that would cause the least contamination and allow the fastest travel based on METT-TC.

**(2)** Tell vehicle operators to close all doors, windows, hatches, and vents on vehicles.

**(3)** Assume MOPP4.

**(4)** Have Soldiers attach M9 detector paper to their clothing and equipment.

**b.** During crossing—

**(1)** Ensure that vehicles are far enough apart during movement to minimize dust. Consider the tactical situation, command, and control when spacing vehicles.

**(2)** Move through the area as quickly as possible. Do NOT forget safety or security. Continuously monitor personnel for chemical-agent symptoms, and give first aid as required.

**(3)** Ensure that Soldiers avoid touching anything in the area if possible.

**(4)** Monitor personnel closely for symptoms of heat stress, and minimize excessive heat buildup.

**c.** After crossing—

**(1)** Continue to monitor Soldiers for chemical-agent symptoms, and give first aid as required.

**(2)** Have Soldiers use detector paper to check themselves and their equipment for contamination.

**(3)** Have Soldiers use their decontaminating kits to decontaminate any contaminated skin or personal equipment. Seek medical aid as required.

**(4)** Have vehicle operators use available decontaminating equipment to decontaminate their vehicles if required.

**3.** Supervise a unit crossing a biological contaminated area.

**a.** Before crossing—

**(1)** Select the shortest possible route that would cause the least contamination and allow the fastest travel based on METT-TC.

**(2)** Tell vehicle operators to close all doors, windows, hatches, and vents on vehicles.

**(3)** Assume the appropriate MOPP level as required.

| Performance Steps |
|---|

**b.** During crossing—

**(1)** Move through the area as quickly as possible. Do NOT forget safety or security.

**(2)** Ensure that personnel do not touch anything in the area that can be avoided.

**(3)** Ensure that vehicles are far enough apart during movement to minimize dust. Consider the tactical situation, command, and control when spacing vehicles.

**c.** After crossing—

**(1)** Decontaminate personnel and equipment by washing them thoroughly with hot, soapy water if time and mission permit.

**(2)** Do as many of the steps as possible for the types of contamination present in the area.

**(3)** Seek medical aid if required.

**Evaluation Preparation:** *Setup*: Evaluate this task during a field exercise or a normal training session. The contaminated area may have been marked with the appropriate markers; or a diagram may have been prepared showing the boundary of the contaminated area, the Soldier's present location, and the desired direction of travel. Gather the necessary equipment and personnel to conduct the movement. Obtain, for training purposes, a fictitious commander's operational exposure guidance (OEG) on the turn-back dose and the turn-back dose rate.

***Note***: Before conducting this task, ensure that Soldiers have been trained on the following tasks: 031-503-1001, 031-503-1037, 031-503-1013, and 031-503-1009.

*Brief Soldier:* Tell the Soldier to supervise the movement of a unit through a CBRN contaminated area by performing the appropriate measures before, during, and after the movement through the area. Give the Soldier an illustration, showing the boundary of the contaminated area, the Soldier's present location, and desired direction of travel through the contaminated area.

| Performance Measures | GO | NO GO |
|---|---|---|
| **1.** Supervised a unit crossing a nuclear contaminated area. | —— | —— |
| **2.** Supervised a unit crossing a chemical contaminated area. | —— | —— |
| **3.** Supervised a unit crossing a biological contaminated area. | —— | —— |

**Evaluation Guidance:** Refer to chapter 1, paragraph 1-4 b (6).

**References**

**Required:** FM 3-11.3, FM 3-11.4, and FM 3-11.5

**Related:**

# 052-195-3066

## Direct Construction of Nonexplosive Obstacles

**Conditions:** You are given a mission directive, logistical support for the type of nonexplosive obstacles, squad personnel, organic equipment, and a FM 5-34.

**Standards:** Direct the construction of nonexplosive obstacles that are tied to existing or reinforced obstacles to block, channelize, or delay an enemy according to the mission brief or intent.

**Performance Steps**

**1.** Analyze mission requirements using the mission, enemy, terrain and weather, troops and support available, time available, civil considerations (METT-TC) factors. Consider the following:

**a.** Manpower available.

**b.** Barrier materials available.

**c.** Time available.

**d.** Type of transportation.

**e.** Site selection (take advantage of existing obstacles).

**f.** Type of obstacle to be constructed.

**g.** Equipment available.

**2.** Direct the construction of wire obstacles.

***Note***: When using U-shaped pickets to construct obstacles, ensure that the open end of the U faces the enemy.

**a.** Direct the construction of a triple-standard concertina fence.

**(1)** Ensure that work is done from the enemy side to the friendly side.

**(2)** Direct the installation of an anchor picket (short picket) at each end of the front row (located on the enemy side) and back row (located on the friendly side) with the first long picket being spaced 1.5 meters (2 paces) from the anchor picket.

**(3)** Ensure that long pickets are spaced at 3.8-meter (5-pace) intervals, staggering the pickets on the back row between the pickets on the front row (figure 052-195-3066-1).

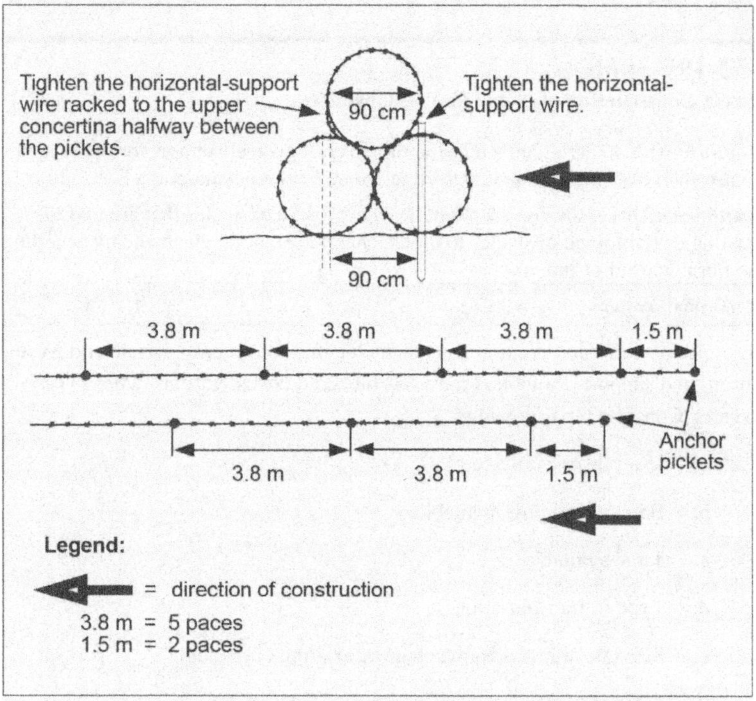


**Figure 052-195-3066-1.Triple-standard concertina fence**

***Note***: Install the pickets in such a manner that the lower notch of the long picket is 10 centimeters (4 inches) aboveground.

**(4)** Direct the installation of the concertina wire. Ensure that—

**(a)** **O**ne roll of concertina wire is placed on the front side at every third picket and two rolls are placed on the back side at every third picket thereafter (figure 052-195-3066-2).

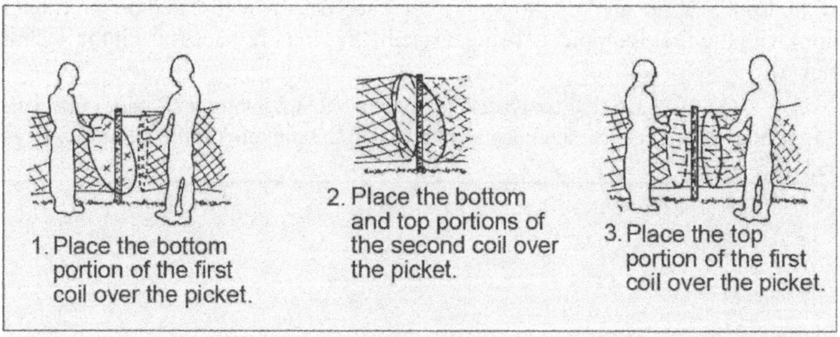


**Figure 052-195-3066-2.Joining the concertina wire**

**Performance Steps**

**(b)** The bottom rows of concertina wire are secured with a horizontal wire (barbed wire) on each row (figure 052-195-3066-3).

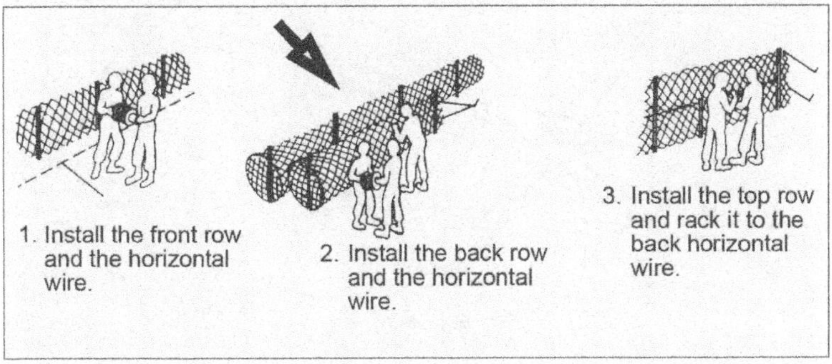


**Figure 052-195-3066-3.Installing the concertina and horizontal wires**

**(c)** The back row is begun once the first team has cleared the head of the first row of the fence.

**(5)** Direct the installation of the remaining row of concertina wire on top of the first two rows, and fasten it securely to the back horizontal wire.

**(6)** Ensure that the concertina wire is properly tied and that all horizontal wires are properly installed.

**b.** Direct the construction of a knife rest.

**(1)** Ensure that work is done from the enemy side to the friendly side and from the bottom up.

**(2)** Ensure that the knife rest is 3 to 5 meters (10 to 16 feet) long (figure 052-195-3066-4). Ensure that—

**(a)** The end poles (1.2 meters [4 feet] in height and width) are lashed to form two Xs.

**(b)** The center pole (3 to 5 meters [10 to 16 feet] in length) is laid across the top intersection of the Xs with each end of the pole lashed to the X.

**(c)** Barbed wire is used to complete the knife rest.

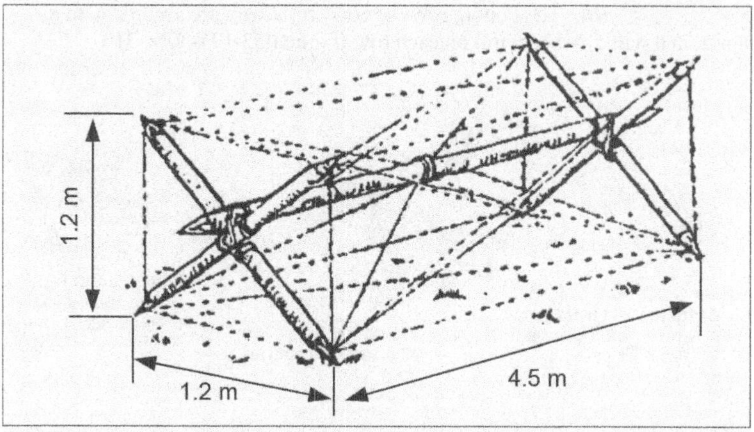


**Figure 052-195-3066-4. Knife rest**

c. Direct the construction of an 11-row obstacle out of concertina wire.

(1) Ensure that work is done from the enemy to the friendly side.

(2) Ensure that long pickets are installed at 3.8-meter (5-pace) intervals for 11 rows (figure 052-195-3066-5).

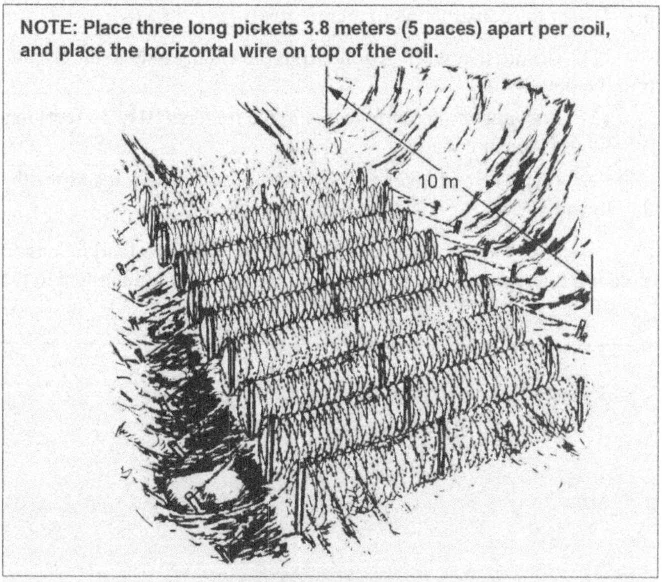


**Figure 052-195-3066-5. An 11-row obstacle**

**(3)** Direct the placement of the concertina wire over the long pickets.

**(4)** Ensure that the horizontal wire (barbed wire) is anchored to the anchor pickets, which is 1.5 meters (2 paces) from each end of the concertina rows.

**(5)** Ensure that the horizontal wire is secured over the row of concertina wire.

**(6)** Ensure that a log (20 centimeters [8 inches] in diameter) is placed between the fifth and sixth rows.

**(7)** Ensure that the obstacle is no less than 10 meters (33 feet) deep.

**3.** Direct the construction of antitank (AT) ditch obstacles (figure 052-195-3066-6).

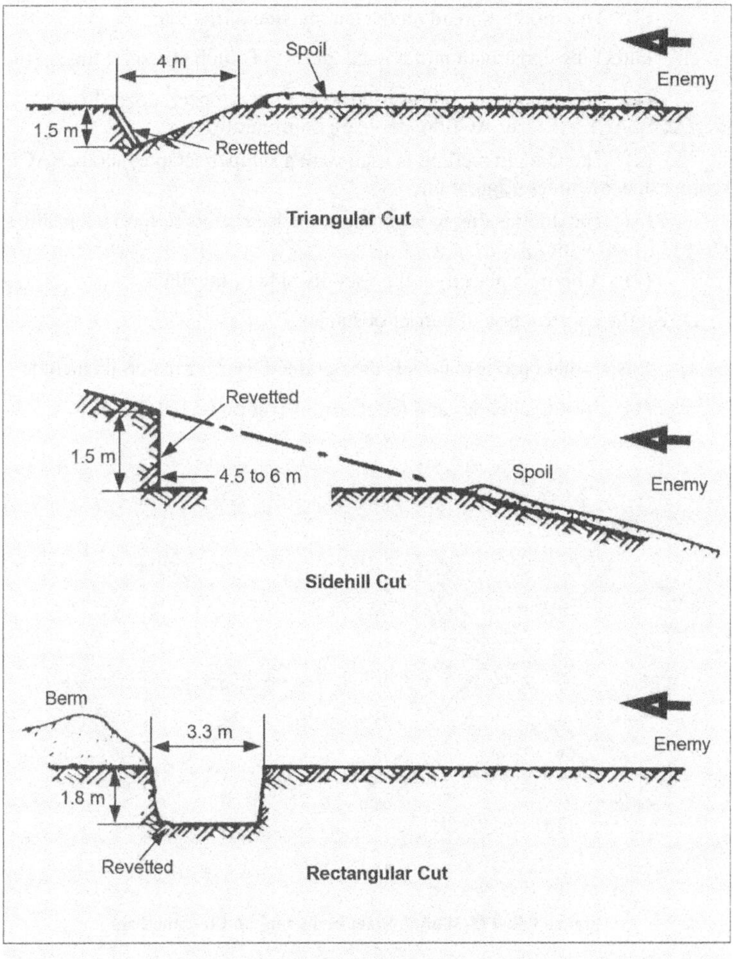


**Figure 052-195-3066-6. AT ditch obstacles**

**Performance Steps**

**a.** Direct the construction a triangular AT ditch obstacle. Ensure that—

**(1)** The T-push method is used with a dozer/dozer, dozer/loader, dozer/armored combat earthmover (ACE), ACE/ACE, or ACE/loader team configuration.

**(2)** The ditch is dug to a minimum of 1.5 meters (5 feet) deep and 4 meters (13 feet) wide.

**(3)** The spoil is spread on the enemy side of the ditch.

**b.** Direct the construction of a sidehill-cut AT ditch obstacle. Ensure that—

**(1)** The T-push method is used with a dozer/dozer, dozer/loader, dozer/ACE, ACE/ACE, or ACE/loader team configuration.

**(2)** The tank ditch is cut to a minimum of 1.5 meters (5 feet) deep and 4.5 to 6 meters (15 to 20 feet) wide.

**(3)** The spoil is spread on the enemy side of the ditch.

**c.** Direct the construction of a rectangular AT ditch obstacle. Ensure that—

**(1)** The T-push method is used with a dozer/dozer, dozer/loader, dozer/ACE, ACE/ACE, or ACE/loader team configuration.

**(2)** The tandem method is used with a scraper/scraper, scraper/ACE, or a scraper/dozer team configuration.

**(3)** The ditch is dug to a minimum of 1.8 meters (6 feet) deep and 3.3 meters (11 feet) wide.

**(4)** A berm is placed on the friendly side of the ditch.

**4.** Direct the construction of barrier obstacles.

**a.** Ensure that barrier materials are obtained. Barrier materials include—

**(1)** Steel hedgehogs and tetrahedrons (figure 052-195-3066-7).

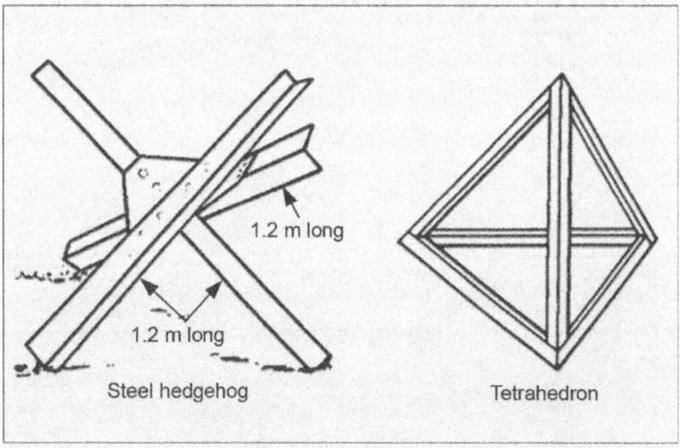


**Figure 052-195-3066-7. Steel hedgehog and tetrahedron**

(2) Concrete tetrahedrons and cubes (figure 052-195-3066-8).

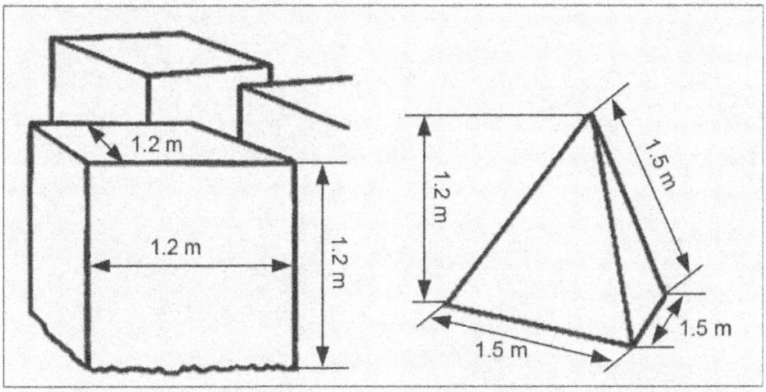


**Figure 052-195-3066-8. Concrete Cubes and tetrahedron**

(3) Jersey barriers (figure 052-195-3066-9).

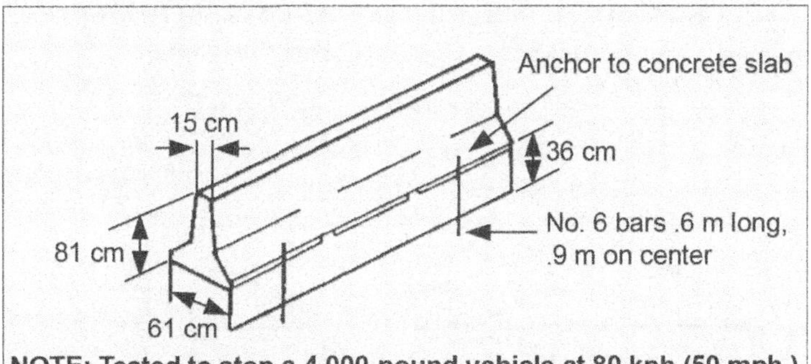


**Figure 052-195-3066-9. Jersey barrier**

**Performance Steps**

**(4)** HESCO Bastion Concertainer® (figure 052-195-3066-10).

**Figure 052-195-3066-10. HESCO Bastion Concertainer®**

**b.** Ensure that barrier materials (figure 052-195-3066-11) are emplaced.

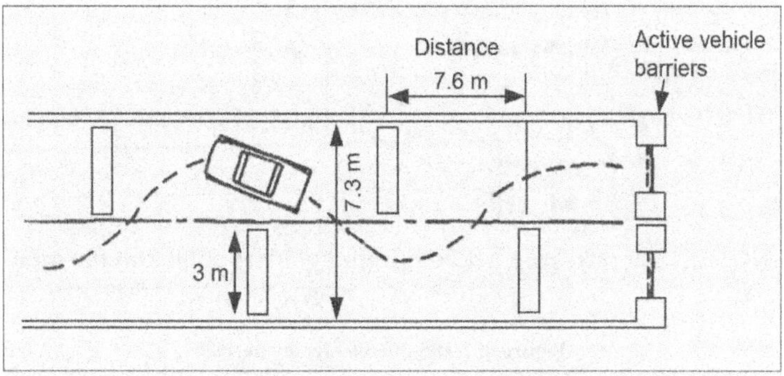


**Figure 052-195-3066-11. Concrete obstacle placement**

***Note***: The distance between barriers will vary depending on the type of traffic expected (cars, trucks, or semitrailers).

**Evaluation Preparation:** *Setup*: Provide the Soldier with the items listed in the conditions and a mission brief describing the type of obstacle to be emplaced.

*Brief Soldier:* Tell the Soldier to direct the construction of the obstacle according to the mission brief.

| Performance Measures | GO | NO GO |
|---|---|---|
| **1.** Analyzed mission requirements using the METT-TC factors. | ___ | ___ |
| **2.** Directed the construction of wire obstacles. | ___ | ___ |
| **3.** Directed the construction of AT ditch obstacles. | ___ | ___ |
| **4.** Directed the construction of barrier obstacles. | ___ | ___ |

**Evaluation Guidance:** Refer to chapter 1, paragraph 1-4 b (6).

**References**

**Required:** FM 5-34

**Related:** FM 5-102

## 071-331-0820

### Analyze Terrain

**Conditions:** Given any tactical mission that involves a specified route or location on the ground or a map, and a standard 1:50,000 scale military map.

**Standards:** Analyze the route or location in terms of the five military aspects of terrain and determine how each aspect affects the mission.

**Performance Steps**

*Note*: To help in analyzing terrain, use the key word OAKOC (observation and fields of fire, avenues of approach, key terrain, obstacles, and cover and concealment).

**1.** Observation and fields of fire.

**a.** Observation requires terrain that permits a force to locate the enemy, either visually or through surveillance devices. The best observation generally is obtained from the highest terrain features in an area. Analyze the effects of visibility on observation with weather rather than terrain, because visibility varies with weather, whereas observation varies with terrain.

**b.** Fire encompasses the influence of the terrain on the effectiveness of direct and indirect fire weapons. Indirect fire is mainly affected by terrain conditions within the target area. Fields of fire for direct fire weapons are mainly affected by terrain conditions between the weapon and target.

**c.** Identify the terrain features in and by the area of operations (AO) that gives the friendly or enemy force favorable observation and fire. Consider these terrain features in your subsequent analysis of key terrain, enemy forces, and cover and concealment.

**Performance Steps**

**2.** Avenues of approach.

**a.** An avenue of approach is a route for a force of a particular size to reach an objective or key terrain. To be an avenue of approach, a route must be wide enough to deploy the size force that will be using it.

**b.** Analyze an avenue of approach solely on the following terrain considerations:

**(1)** Observation and fire. Determine if the avenue of approach provides favorable observation and fire for the force moving on it.

**(2)** Concealment and cover. Determine if the avenue of approach provides cover and concealment. Both can conflict with observation and fire.

**(3)** Obstacles. Determine if the avenue of approach avoids obstacles that are perpendicular to the direction of advance and, when practical, that takes advantage of those that are parallel to the direction of advance.

**(4)** Use of key terrain.

**(5)** Adequate maneuver space.

**(6)** Ease of movement.

**3.** Key terrain. A key terrain feature is any point or area that seizure or control affords a marked advantage to either force. "Seizure" means physical occupation of the terrain by a force, whereas "control" might or might not include physical occupation. The selection of key terrain varies with the level of command, the type of unit, and the unit's mission.

**4.** Obstacles.

**a.** An obstacle is any natural or artificial terrain feature that stops or impedes military movement.

**b.** The mission influences consideration of obstacles.

**c.** An obstacle might be an advantage or disadvantage. Consider each on its own merits, and for each specific mission. For example, obstacles perpendicular to a direction of attack favor the defender because they slow or channelize the attacker. Obstacles parallel to the direction of attack can help protect the flank of the attacking force.

**5.** Cover and concealment. Cover is protection from the effects of fire. Concealment is protection from observation. You must determine the cover and concealment available to both friendly and enemy forces.

**a.** Cover might be provided by terrain features or manmade features. Areas that provide cover from direct fire might or might not protect against the effects of indirect fire. Most terrain features that offer cover also offer concealment.

**b.** Concealment might be provided by terrain features, vegetation (such as wood, underbrush, or cultivated vegetation), or any other feature that denies observation. Concealment does not necessarily provide cover.

**Evaluation Preparation:** *Setup*: Give the Soldier a tactical mission that involves a specified route or location on the ground or a map. If only a map is used, issue a standard 1:50,000 scale military map to the Soldier.

*Brief Soldier:* Tell the Soldier to analyze the route or location in terms of the five military aspects of terrain and determine how each aspect affects the mission.

| Performance Measures | GO | NO GO |
|---|---|---|
| **1.** Identified effects of terrain on observation and fire. | —— | —— |
| **2.** Identified concealment and cover along the route or at the position. | —— | —— |
| **3.** Identified obstacles that would be an advantage and a disadvantage along the route or at the position. | —— | —— |
| **4.** Identified key terrain along the route or at the position. | —— | —— |
| **5.** Identified the best avenue of approach. | —— | —— |

**Evaluation Guidance:** Refer to chapter 1, paragraph 1-4 b (6).

**References**

**Required:**

**Related:**

## 071-410-0012

### Conduct Occupation of an Assembly Area

**Conditions:** As a platoon-sized unit leader, given a company commander's order to occupy a specific sector of a company assembly area and a map of the operational area.

**Standards:** Move the unit to the location specified in the company order; ensure preparation of the assigned sector is completed in the time specified in the order; and position elements, weapons, and observation posts (OPs) in a manner that support the company plan for occupying the assembly area (AA).

| Performance Steps |
|---|

**1.** Prepare to occupy the AA.

**a.** Perform the required troop-leading procedures (TLPs). TLPs are a series of eight interrelated, overlapping processes that are often accomplished concurrently and do not follow a rigid sequence. Use the procedures as outlined below, if only in abbreviated form, to ensure that nothing is left out of planning and preparation and that Soldiers understand the mission and prepare adequately.

**(1)** Receive the mission in the company operation order (OPORD).

**(2)** Issue the warning order to subordinate leaders. Include location, special equipment required, and the earliest time for movement.

**Performance Steps**

**(3)** Make a tentative plan for moving to and preparing the position based on the estimate of the situation and an analysis of mission, enemy, terrain and weather, troops and support available, time available, civil considerations (METT-TC). Your plan must support the company plan for occupying the AA.

**(4)** Start necessary movement. Movement might need to start while you are still planning. Movement can occur anytime during the TLP.

**(5)** Reconnoiter the position and the route(s) to it. The situation might prohibit this. At least conduct a map reconnaissance to confirm or deny assumptions made during the estimate of the situation.

**(6)** Complete the plan.

**(7)** Issue the order to subordinate leaders. Use the standard OPORD format. At a minimum, include—

**(a)** Situation.

**(b)** Mission and purpose for occupying the AA.

**(c)** Each squad's position (left, right, center) in the platoon sector.

**(d)** Security plan (passwords, OPs, percent of personnel on alert).

**(e)** Times for movement or occupation of the AA.

**(f)** Other pertinent information such as location of the command post, methods of waste disposal, and plan for dealing with environmental hazards.

**(8)** Supervise continuously.

**b.** As needed, coordinate with elements that will be attached to or adjacent to your position in the AA.

**c.** Provide platoon representatives for the company quartering party. Tell them to perform the following tasks or other tasks as required:

**(1)** Reconnoiter the AA to ensure that it is clear of the enemy.
**(2)** Establish initial security.
**(3)** Select initial positions for all platoon elements.
**(4)** Identify, clear, or mark obstacles in the platoon sector of the AA.

**2.** Direct platoon movement to the designated platoon release point. Use appropriate movement techniques based on the terrain and the situation.

**3.** Prepare your sector of the AA according to the company plan.

**a.** Link up with guides and move the platoon to its initial position.

**b.** Establish and maintain local security.

**c.** Assign squad sectors. Ensure that sectors are mutually supporting and that all gaps are covered by fire and observation.

**d.** Designate OP(s) locations and the elements responsible for establishing and maintaining them.

**Performance Steps**

**e.** Ensure that communications are established within the platoon and company.

**f.** Submit timely progress reports to company headquarters.

**g.** Establish and enforce priority of work. The following is an example of work priority and may vary based on the unit standing operating procedure (SOP), mission, or METT-TC:

**(1)** Position vehicles, crew-served weapons, and chemical-agent alarms, and designate principal direction of fire (PDF), final protective line (FPL), and final protective fire (FPF).

**(2)** Construct fighting position.

**(3)** Set up wire communications.

**(4)** Prepare range cards.

**(5)** Distribute ammunition, rations, water, supplies, and special equipment.

**(6)** Conduct preventive maintenance checks and services (PMCS) on equipment.

**(7)** Inspect personnel and equipment.

**(8)** Rehearse critical aspects of the upcoming mission.

**(9)** Test small arms (if situation permits).

**(10)** Conduct personal hygiene and field sanitation.

**(11)** Institute a rest plan.

**h.** Coordinate with adjacent units and others as required.

**(1)** Coordinate for security patrols (if applicable).

**(2)** Establish responsibility for overlapping enemy avenues of approach between adjacent units.

**(3)** Ensure that there are no gaps between elements.

**(4)** Exchange information on OP locations and unit signals.

**(5)** Coordinate for local counterattacks.

**(6)** Complete and forward a copy of the platoon sector sketch to company headquarters.

**Evaluation Preparation:** *Setup*: This task should be evaluated during a field training exercise. Otherwise, assign an assembly area to be occupied and a quartering party that has accomplished its tasks according to the company and platoon SOP.

*Brief Soldier:* Issue an OPORD for an operation requiring the occupation of an assembly area.

| Performance Measures | GO | NO GO |
|---|---|---|
| **1.** Conducted preparations for occupying the AA. | —— | —— |
| **2.** Directed platoon movement to the designated release point. Used the appropriate movement technique based on the terrain and situation. | —— | —— |

| Performance Measures | GO | NO GO |
|---|---|---|
| **3.** Prepared assigned sector of AA according to the company plan. | ___ | ___ |

**Evaluation Guidance:** Refer to chapter 1, paragraph 1-4 b (6).

**References**

**Required:**

**Related:** FM 3-21.8 (FM 7-8), FM 3-21.71, and FM 7-7

---

## 071-420-0021

### Conduct a Movement to Contact by a Platoon

**Conditions:** In a combat environment, given a platoon and a mission requiring movement to contact.

**Standards:**

1. Develop a plan that includes objectives, routes, key terrain, fire support, formations, and other required control measures.

2. Include in the preparations the cleaning and test firing of weapons, maintenance of vehicles and equipment, and the resupply of ammunition or needed equipment.

3. Make contact with the threat with the smallest possible element. Develop the situation by controlling movement and fires. Submit situation reports as required.

| Performance Steps |
|---|

**1.** Receive and analyze the mission.

*Note*: A warning order should be issued to provide subordinates with as much preparation time as possible.

**2.** Make a tentative plan.

**3.** Start preparations.

**4.** Conduct reconnaissance.

**5.** Finalize the plan.

*Note*: Preparation continues as the plan develops. New information is incorporated into preparations. The operation order is issued at the time and place specified in the warning order.

**6.** Continue preparations.

**7.** Request supplies.

**8.** Receive supplies.

**9.** Conduct rehearsals.

**10.** Conduct final inspection.

**11.** Control movement.

| Performance Steps |
|---|
| **12.** Control fires. |
| **13.** Report contact to the company. |

**Evaluation Preparation:** *Setup*: At the test site, provide an area in which a movement to contact can be conducted. Test this task during a platoon or larger tactical exercise.

*Brief Soldier*: Tell the Soldier, as the leader of a platoon, to conduct a movement to contact and then must react properly to enemy fire.

| Performance Measures | GO | NO GO |
|---|---|---|
| **1.** Received and analyzed the mission. | —— | —— |
| **2.** Made a tentative plan. | —— | —— |
| **3.** Started the preparations. | —— | —— |
| **4.** Conducted the reconnaissance. | —— | —— |
| **5.** Finalized the plan. | —— | —— |
| **6.** Continued preparations. | —— | —— |
| **7.** Requested supplies. | —— | —— |
| **8.** Received supplies. | —— | —— |
| **9.** Conducted rehearsals. | —— | —— |
| **10.** Conducted the final inspection. | —— | —— |
| **11.** Controlled movement. | —— | —— |
| **12.** Controlled fires. | —— | —— |
| **13.** Reported contact to the company. | —— | —— |

**Evaluation Guidance:** Refer to chapter 1, paragraph 1-4 b (6).

**References**

**Required:**

**Related:** FM 3-21.8 (FM 7-8), FM 3-21.71, and FM 7-7

## SUBJECT AREA 5: NAVIGATE

# 071-332-5000

## Prepare an Operation Overlay

**Conditions:** Given a complete copy of the operation order (OPORD) that your unit is to execute, a commander's or a battalion operations officer's (S3) guidance (to include time available for preparation), overlay paper, tape, a map of the operational area, colored pencils (red, black, blue, green, and yellow), a No. 2 pencil, a coordinate scale, and FM 1-02 (FM 101-5-1).

**Standards:** Within the time specified, ensure that the overlay—

1. Is identified by map reference data, effective date, and purpose.
2. Contains classification markings and downgrading instructions, if applicable.
3. Contains distribution instructions and authentication, if distributed separately.
4. Is prepared according to overlay techniques outlined in FM 1-02 (FM 101-5-1).
5. Is prepared with boundaries and unit locations plotted to within 50 meters and tactics and fire support measures to within 25 meters.

**Performance Steps**

*Note*: Overlay techniques involve the use of military symbols to portray, in a condensed form, the plans, orders, and information concerning a military operation.

**1.** Register the overlay.

**a.** Orient the overlay material over the map area to the annotated portion and temporarily attach edit to the map with the tape.

**b.** Trace the grid intersections nearest the opposite corners of the overlay and label each with the proper grid coordinates.

**2.** Plot detail. Use colored pencils or markers in standard colors, when available, to plot any detail (FM 1-02); otherwise, plot the activity to be shown with a pen or pencil that makes a lasting mark, but without cutting the overlay. Use standard military symbols where possible. When nonstandard symbols are used, identify them on the edge of the overlay. Show only the detail with which the document is directly concerned. Use the following standard colors:

**a.** Blue or black: Friendly units, installations, equipment, and activities.

**b.** Red: Enemy installations, equipment, and activities.

**c.** Yellow: Any areas of chemical, biological or radiological contamination.

**d.** Green: Any manmade obstacle.

*Note*. If only one color is available, show enemy symbols with double lines.

**3.** Mark the overlay's classification. The classification depends on the classification of the information used to prepare the overlay. Mark the classification on the top and bottom of the overlay.

**4.** Use the following overlay techniques:

**a.** Solid and broken lines.

**(1)** Use solid lines to represent the location of a unit or installation or coordinating detail (such as a line of departure or boundary) that is in effect and will continue; or a location that the operation order will affect.

(2) Use broken lines to represent any proposed or future location, or any coordinating detail to become effective later:

**b.** Indicated boundaries.

**(1)** Boundaries show the areas of tactical responsibility. In the offense, these are called zones of action. In the defense and retrograde, they are called sectors of responsibility. When you describe them verbally, describe their lateral boundaries from rear to front in the offense, and from front to rear in the defense and retrograde.

**(2)** Use rear boundaries when you must precisely define the area of responsibility for forward units. When no rear boundary is delineated, determine the rear limit of a unit's area of responsibility by visualizing a rear boundary, drawn generally parallel to the front, along a natural terrain feature, and connecting at the rearward limit of the unit's lateral boundaries.

**(3)** Mark the rear boundary as shown. Ensure that the size shown along the boundary corresponds to the low unit. Show the arm or branch as needed to prevent confusion (figure 071-332-5000-1).

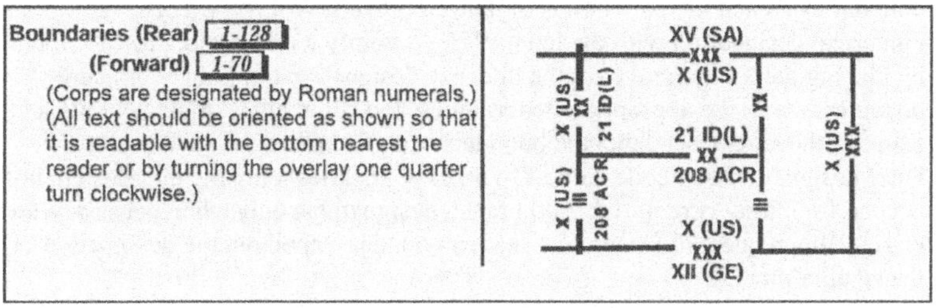


**Figure 071-332-5000-1. Rear boundary**

**(4)** Draw boundaries along terrain features that are easily recognizable on the ground, and where situated, if possible, so that the key terrain features, avenues of approach, and river can completely include the one unit. Show them with a solid line if they are currently in effect or if the OPORD will affect them. Using them is based on the techniques and tactics peculiar to the type of tactical operation.

**(5)** Use a broken line to represent future or proposed boundaries, and label them to indicate their effective time, if appropriate (figure 071-332-5000-2).

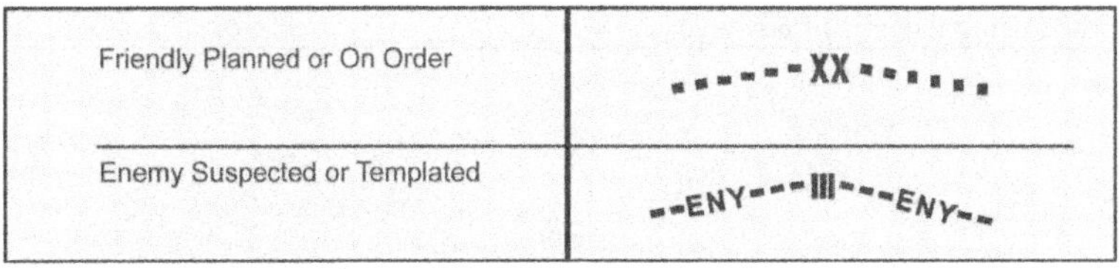


**Figure 071-332-5000-2. Proposed boundary**

**(6)** Mark boundaries as follows: Place a symbol on the boundary to show size and designation of the highest units that shared the boundary.

**(7)** Show units of unequal size. Completely show the symbols of the higher unit and the designation of the lower unit to show its size. The boundary between the 52ID (M) and the 312 SIB (M) is shown in figure 071-332-5000-3.

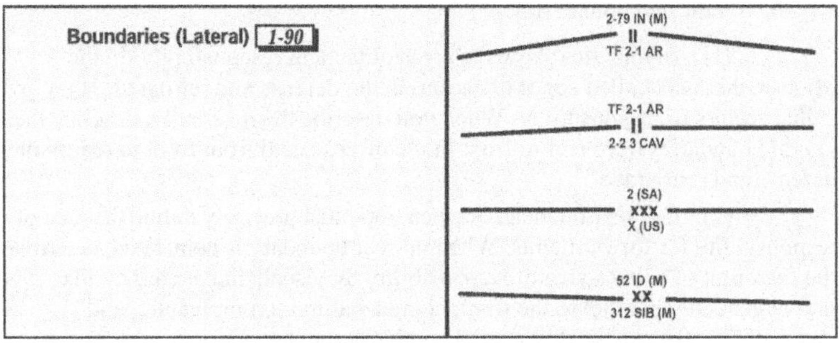


**Figure 071-332-5000-3. Lateral boundary**

**(8)** Identify overlays or sketches accompanying written or oral orders that specify task organization. Indicate numerical unit designation on battalion boundaries. When the battalion is organized into a task force, precede the numerical designation with the letters "TF." Identify a unit symbol as task force by placing the TF symbol over the unit size designation (||). Label company boundaries with the appropriate letter, unless the company is organized into a team. In the latter case, label the boundary with the abbreviation TM plus the letter designation or a code name. On other boundaries, only the unit designation needed for clarity is required. Add branch designations only when necessary for clarity. When unequal-sized units share a boundary, spell out the designation of the smaller unit.

**c.** Marked axis of advance.

**(1)** Extend an axis of advance arrow only as far as this form of control is essential to the overall plan. Normally, it is shown from the line of departure (LD) to the objective following an avenue of approach. It shows that the commanders may maneuver their forces and place them freely to either side of the axis. The purpose is to avoid obstacles, engage the enemy, or bypass enemy forces that cannot threaten security or jeopardize accomplishing the mission. The commander ensures that such deviation does not interfere with adjacent units, that the unit remains oriented on the objective, and that the location and size of the bypassed enemy forces is reported to higher headquarters. When the situation dictates, the commander assigns boundaries as an additional control measure when using the axis of advance.

**(2)** Show an axis of advance as shown below, identified by a code. You could identify it with a unit designation (figure 071-332-5000-4).

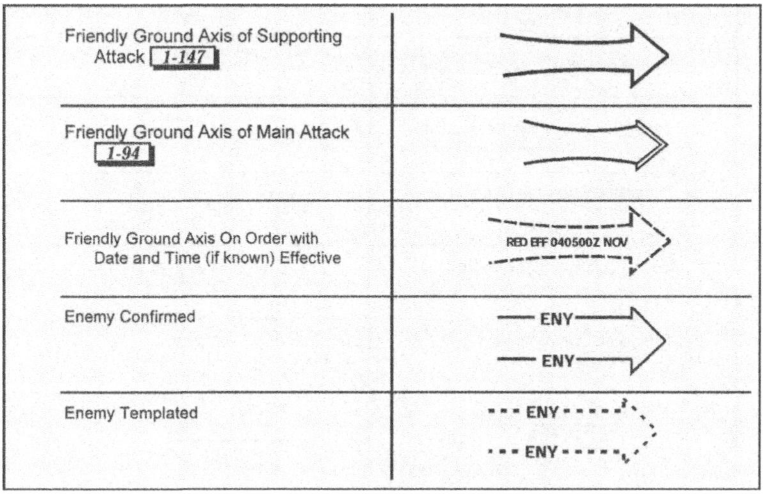


**Figure 071-332-5000-4. Axis of advance**

**(3)** Differentiate between a ground axis of advance and an air assault of advance. To do so, use a twist in the shaft of the open arrow, like a propeller (figure 071-332-5000-5).

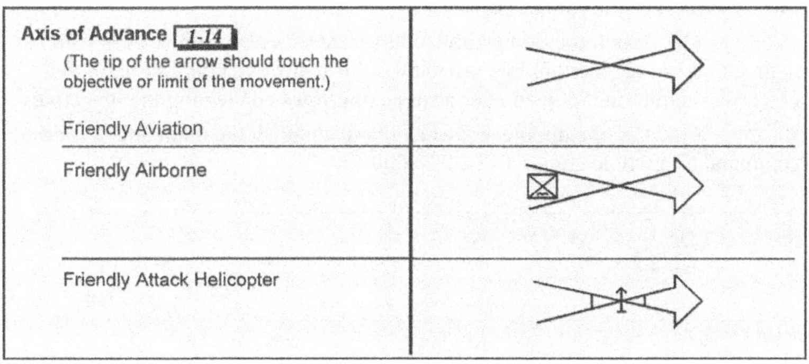


**Figure071-332-5000-5. Air assault axis of advance**

**d.** Show direction of attack arrows. Use this control measure when the commander wants to say where the center of mass of a subordinate unit must move in an attack to ensure the accomplishment of a closely coordinated plan of maneuver. One example would be a night attack or counterattack. Extend a direction of attack arrow from the LD to the objective. Do not label it (figure 071-332-5000-6).

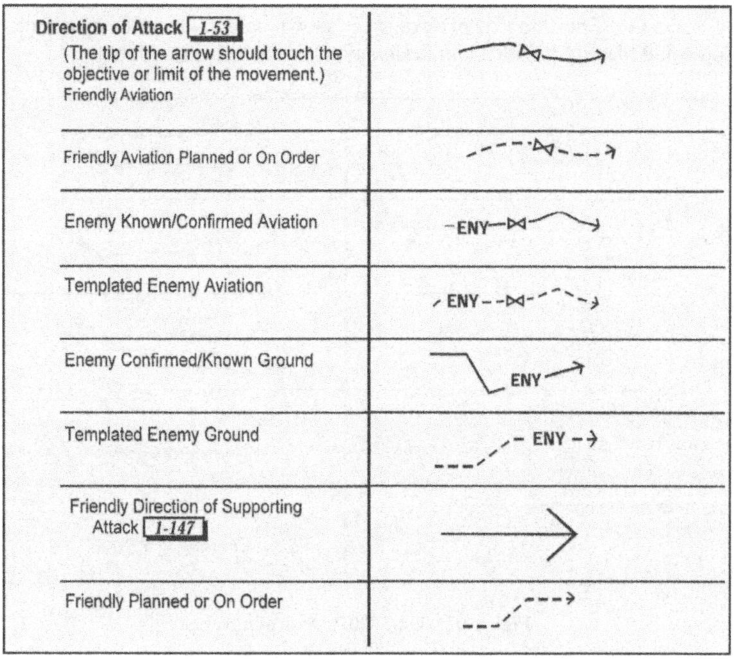


**Figure 071-332-5000-6. Direction of attack arrow**

**(1)** Use the arrow only where necessary because it restricts the maneuver of the subordinate unit.

**(2)** When the unit is directed to seize successive objectives with its main attack along a certain line, use either one arrow extending through the objectives to the final objective or a series of arrows connecting the objectives.

**(3)** Use the double arrowhead to distinguish the main attack for the command as a whole (figure 071-332-5000-7).

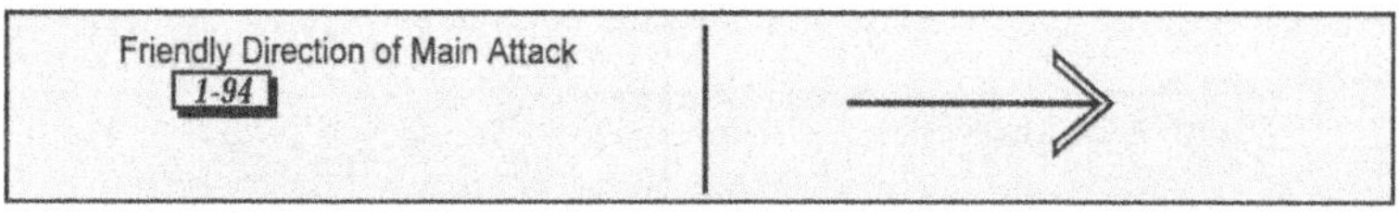


**Figure 071-332-5000-7. Arrowhead indicating the main attack**

**e.** Graphically portray units assigned with a security mission.

**(1)** Show the general location of a unit with a security mission; arrows generally indicate the terrain where the unit operates and the farthest extension of its mission (figure 071-332-5000-8).

Performance Steps

| Note: The letter "S" is used for screening mission. "G" is used for guard mission. "C" is used for covering force. | 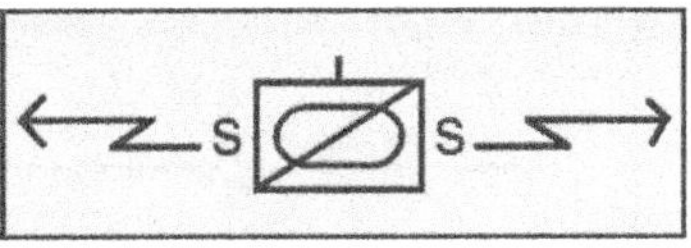 |
|---|---|

Figure 071-332-5000-8. Unit with security mission

(2) Show the primary security mission.

f. Graphically portray supply routes.

(1) The main supply route (MSR) is the route(s) designated within an area of operations where most of the traffic flows in support of the operation. Label the route MSR and assign it a code name.

*Note*: Do not use the term MSR below division level.

(a) In the defense, the division extends the MSR forward to the brigade trains. The brigade's supply route (SR) extends from the battalion trains to a point at the rear of the forward company defense sectors.

(b) In the offense, the proposed SR is shown forward to the objective or as far as the battalion supply officer (S4) can visualize sustainment for the operation. Forward of the LD, show it as a broken line.

(2) Show the symbols to show the division (offensive action) as shown in figure 071-332-5000-9.

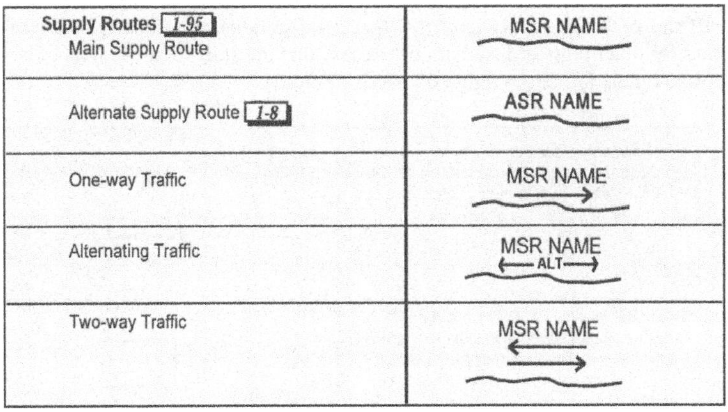


Figure 071-332-5000-9. Divisions MSR

(3) Show sustainment facilities on the operation overlay, or the S4 disseminates their location, as appropriate.

g. Portray unit locations.

(1) To show the location of a unit on an overlay, draw the symbol so that its center corresponds with the coordinates where the unit is located (figure 071-332-5000-10).

## Performance Steps

| | |
|---|---|
| A solid line symbol represents a present or actual location. |  |
| A broken line symbol indicates a future or projected location. | 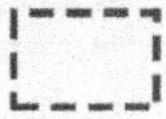 |

**Figure 071-332-5000-10. Location of a unit**

**(2)** Show the location of a trains area, observation posts, or logistical activity. Ensure that the center of the symbol corresponds with the element's coordinates. Figure 071-332-5000-11 shows the location of an observation post.

| | |
|---|---|
| Observation Post/Outpost *1-112* | 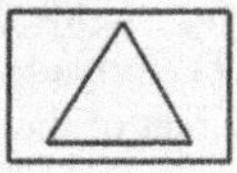 |

**Figure 071-332-5000-11. Location of an observation post**

**(3)** Use the offset technique for clarity when space precludes normal placement of symbols. "Bend" offset staffs as required. Dash the offset staff for future or proposed locations. Extend offset staffs vertically from the bottom center of the symbol (except for command posts [CPs]). The end of the offset staff shows the exact locations of CPs and aid stations, as well as the center of mass for other units or installations. Ensure that the staff for a CP symbol is always on the left edge (figure 071-332-5000-12).

| | |
|---|---|
| Basic symbols other than the headquarters symbol (for example, points) may be placed on a staff which is extended or bent. The end of the staff indicates the precise location. |  |
| Since the headquarters symbol already includes a staff, this staff may be extended or bent. The end of the staff, or extension (if used), indicates the exact location of the headquarters. | 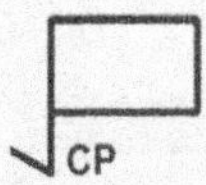 |

**Figure 071-332-5000-12. Offset technique**

**(4)** Locate units.

**(a)** Show the locations of attacking units with boundaries (and CP symbols, when the locations of the CPs are known) or with unit symbols.

**(b)** Show the location of the reserve with an assembly area symbol and with a CP or unit symbol.

Performance Steps

(c) Normally show the reserve units of a force assigned to a defense position or battle position with a line enclosing the area occupied or to be occupied—in other words, a "goose egg." Number or letter these positions for convenience. Figure 071-332-5000-13 shows an occupied and unoccupied company assembly area (reserve location). Figure 071-332-5000-14 shows an occupied and unoccupied reserve company battle position.

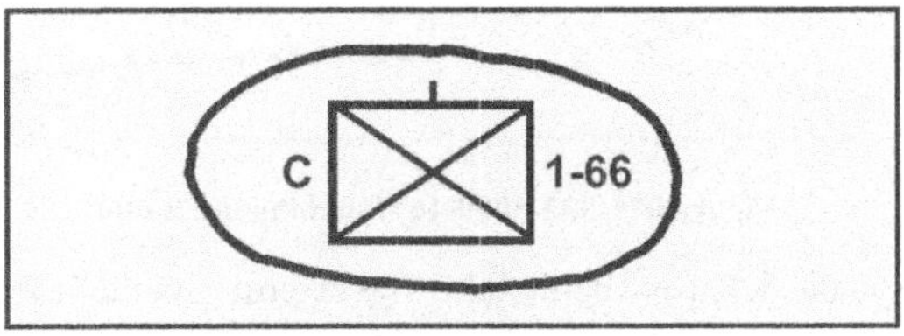


**Figure 071-332-5000-13. Occupied and unoccupied company assembly area (reserve location)**

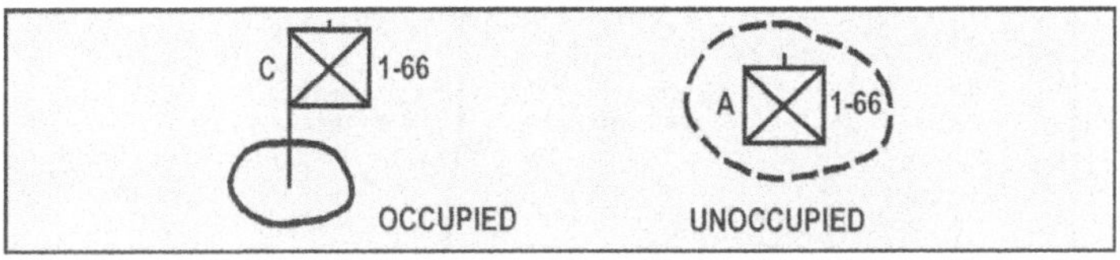


**Figure 071-332-5000-14. Occupied and unoccupied reserve company battle position**

**h.** Identify objective(s). Identify each objective by the abbreviation "OBJ" and a number, letter, or name designation (figure 071-332-5000-15).

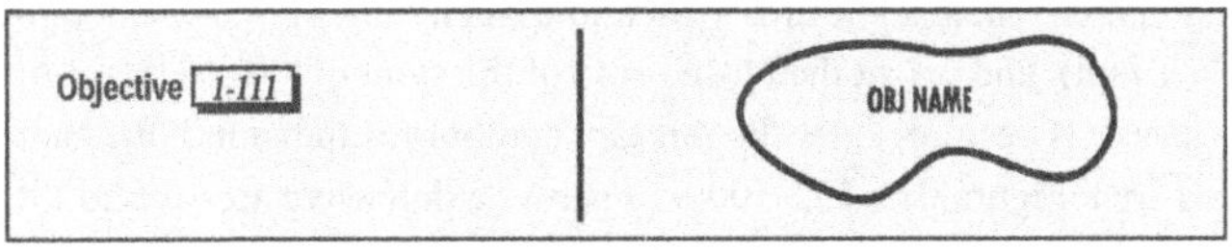


**Figure 071-332-5000-15. Objective**

***Note***: An objective assigned by higher headquarters may be given entirely to one subordinate unit or may be divided. If divided, the objective may be shown graphically as separate objectives and numbered accordingly, or they may be divided into two objectives by a boundary line.

**i.** Pinch out a unit.

**(1)** Show this type of operation by drawing the boundary across the front of the unit, usually along a well-defined terrain feature such as a stream, ridge, or highway.

**(2)** The following example shows that Company A will be pinched out after seizing OBJ 1; then, Company B will seize OBJ 2 and continue the attack to seize OBJ 3 (figure 071-332-5000-16).

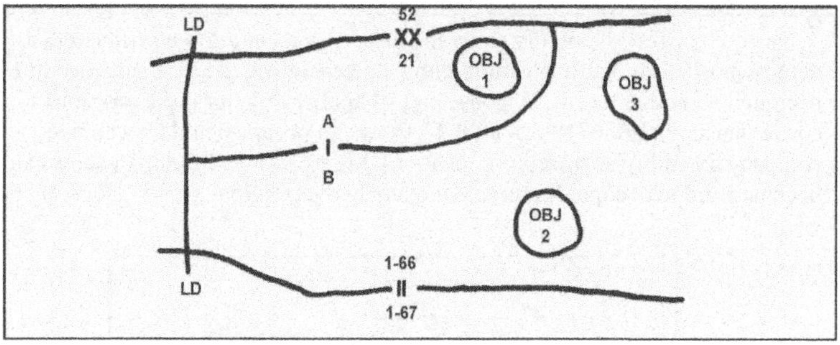


**Figure 071-332-5000-16. Pinching out a unit**

**j.** Show the defensive battlefield. The defensive battlefield is organized into the covering force area and the main battle area (MBA) (figure 071-332-5000-17).

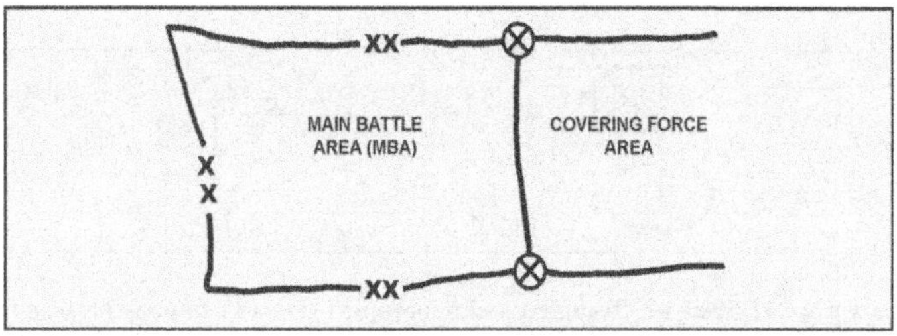


**Figure 071-332-5000-17. Organizing the defensive battlefield**

**k.** Show the defended areas. If an area is occupied and the defense of the area is prepared, enclose the area with a line (including the size symbol of the defending unit), and orient the closed side of the symbol toward the most likely enemy threat. If desired, enter the military symbols of the unit in the center of the enclosed area. Figure 071-332-5000-18 shows a defensive area for 2d Plt, C Co, 1st Bn, 6th Inf, and a proposed defensive area for B Co, 3d Bn, 52d Inf.

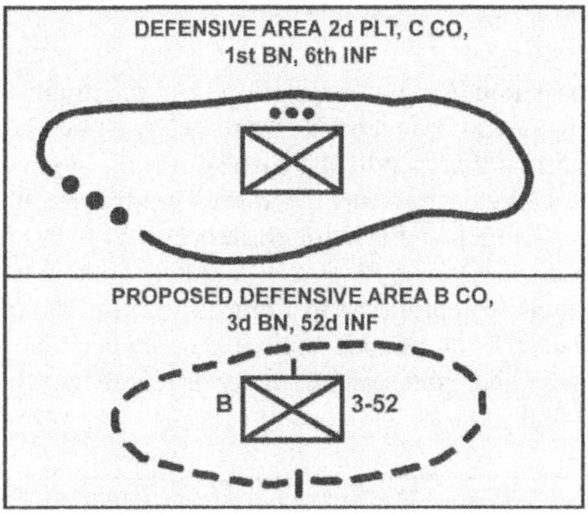


**Figure 071-332-5000-18. Defensive area**

l. Show control measures as required.

**(1)** Line of departure (LD). The LD is a control measure to coordinate the advance of an attacking unit (figure 071-332-5000-19). The LD should be—

**(a)** Clearly defined on the ground and on the map.

**(b)** Roughly perpendicular to the direction of the attack.

**(c)** Under control of friendly units.

**(d)** Marked on both ends.

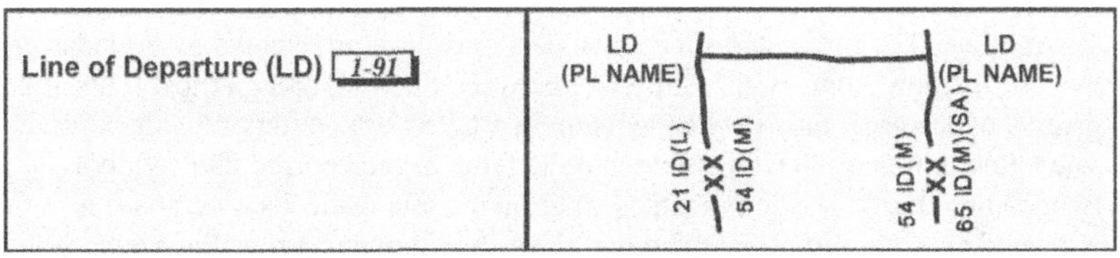


**Figure 071-332-5000-19. Line of departure**

**(2)** Line of contact (LC) (figure 071-332-5000-20).

**Figure 071-332-5000-20. Line of contact**

**(a)** When units are in contact with the enemy, show the frontline as a series of arcs, and label the ends of the arc "LC."

Performance Steps

(b) If the LC is used as an LD, mark it "LD/LC."

(c) If the LC is not used, show the LD with a solid line marked "LD."

(3) Phase line (PLs). Use PLs to control the progress of units for reference in issuing orders or receiving reports. They should be easily recognized terrain features, normally perpendicular to the direction of advance. Also use a PL to control fires and unit movement, and even to limit the advance of attacking elements. Units report their arrival at or clearance of a phase line, but they halt only if ordered to do so. Draw a PL as a solid line with the letters "PL" at each end of the line, or where appropriate to allow easy identification. Identify a PL further by a number, a letter, or code name (using phonetic letters, colors, flowers, cars, or any other code system) under or beside the PL abbreviation (figure 071-332-5000-21).

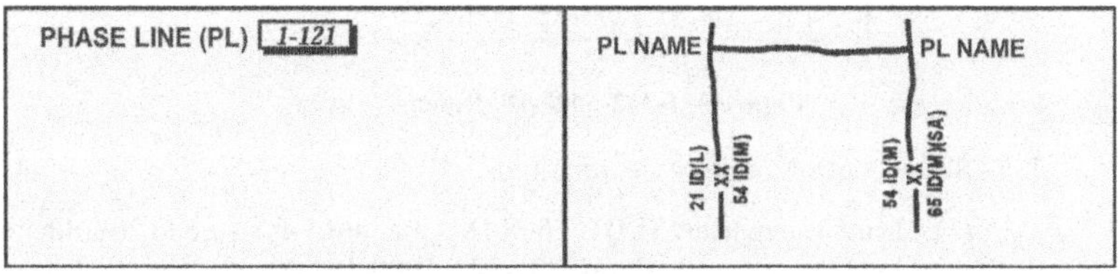


**Figure 071-332-5000-21. Phase line**

(4) Initial delay position (IDP). An IDP is where a delaying action begins, trading space for time. The delay sector is defined by boundaries. The initial and all subsequent delay positions can be related to a time-phased forward edge of the battle area (FEBA). The initial and subsequent delay positions are specified, and PLs may be used to report the progress of the battle. The enemy is held forward of delay lines until the specified time or until permission is granted to withdraw. The initial and successive delay positions are shown on boundaries by coordination points with a solid line between them. Although most IDPs are given a code name, they may have a number, letter, or a variety of code names. The letter abbreviation (IDP) can be to the flank of the coordination symbol (when at the flank, it is in parentheses) or on the line itself. Its time phase is indicated as a date-time group having a two-digit day and a four-digit hour, both connected. The month indicator can be a three-letter type or spelled out, depending upon the desires of the commander. Place the letters "IDP" in parentheses between the line code name, letter, or number and the date-time group (figure 071-332-5000-22).

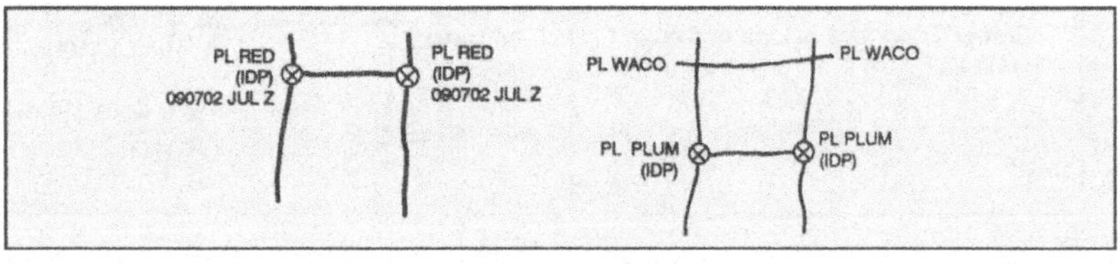


**Figure 071-332-5000-22. Initial delay position**

| Performance Steps |
|---|

(5) Delay lines (DLs). These show the location of a succeeding delay position. Draw delay positions (other than initial) the same, but place the letter abbreviation along the line, and none are placed to the flanks at the coordinating points.

(6) Coordinating points.

(a) Coordinating points are designated on boundaries as specific points for coordination of fires and maneuver between adjacent units. They are indicated whenever a boundary crosses the FEBA and should be indicated whenever the boundary crosses the covering force. Coordinating points are also used where DLs and internal boundaries intersect.

(b) Coordinating points should be located at some terrain feature easily recognizable both on the ground and on a map. Their location on a boundary indicates the general trace of the FEBA, covering force, or DL as visualized by the commander who designates them.

(c) Show the symbol for a coordinating point by a circle with an "X" centered in it (figure 071-332-5000-23). Label the symbol as appropriate.

| | |
|---|---|
| **Coordinating Point** *1-39*<br>**(Center of symbol is exact location.)** |  |

**Figure 071-332-5000-23. Coordinating point**

(7) Checkpoints. Checkpoints, figure 071-332-5000-24, are shown graphically by a number, letter, or code word inside an upright rectangle with a pointed bottom. They are easily recognizable terrain features or objects, such as crossroads, churches, lone buildings, stream junctions, hills, bridges, and railroad crossings. They may be selected throughout the area of operation. By referring to these points, the subordinate commander can quickly and accurately report his/her location. Higher headquarters can use it to designate objectives, boundaries, assembly areas, phase lines, and so forth.

| | |
|---|---|
| **Checkpoint (CKP)** *1-25*<br>**(Bottom point or an extension is exact location.)** |  |

**Figure 071-332-5000-24. Checkpoint**

(8) Contact points. Contact points, shown graphically by squares with a staff extending from the bottom center, are designated at the units to make physical contact. Contact points may also be used to delineate areas of responsibility in specific localities when boundaries are obviously unsuitable such as between elements of a flank guard (figure 071-332-5000-25).

| Contact Point 1-37 |  |
|---|---|



**Figure 071-332-5000-25. Contact point**

**(9)** Passage points (PPs). Show a passage point much like a checkpoint, with the letters "PP" with the number or letter of the passage point within the symbol. Place them along the LD or FEBA of the unit being passed through. The PPs will show where the commander wants the subordinate units to pass.

**(10)** Linkup points. A linkup point should be an easily identifiable point on the ground and map. It is used to facilitate the joining, connecting, or reconnecting of elements of a unit or units. They are used when two or more Army elements are to join each other, when Army and sister service elements are to join each other, and when Army or sister service and allied elements are to join each other. The linkup is an operation in itself and is conducted as part of an airborne or airmobile operation, an attack to assist in the breakout of an airborne or airmobile operation, an attack to assist in the breakout of an encircled force, or an attack to join an infiltrating force. The battalion may participate in a linkup as part of a larger force, or it may, itself, conduct a linkup. The symbol for a linkup point is similar to the one for a checkpoint, but with a dot in the center. Place a number, name, or code name near the symbol to ensure that it is clear and that it refers to that symbol (figure 071-332-5000-26).

| Linkup Point 1-92 |  |
|---|---|



**Figure 071-332-5000-26. Linkup point**

**(11)** Points of departure (PDs). These normally are shown along the LD for night attacks. They look like a checkpoint but contain the letters "PD" and a letter or number. The point of the arrow is at the bottom of its location.

**Evaluation Preparation:** *Setup*: Provide the Soldier with all the material and equipment listed in the condition statement.

*Brief Soldier:* Tell the Soldier to prepare an overlay for the operation indicated in the OPORD.

| Performance Measures | GO | NO GO |
|---|---|---|
| **1.** Placed registering marks in opposite corners of the overlay. | —— | —— |
| **2.** Marked information using standard military symbols. | —— | —— |
| **3.** Indicated enemy installations, equipment, or activities with either the color RED or with double-lined symbols. | —— | —— |

| Performance Measures | GO | NO GO |
|---|---|---|
| **4.** Indicated future locations or coordinating detail with broken-line symbols. | ___ | ___ |
| **5.** Placed symbols indicating size on a boundary to indicate the highest unit sharing the boundary. | ___ | ___ |
| **6.** Extended the axis-of-advance arrow from the LD to the objective. | ___ | ___ |
| **7.** Labeled the route-of-march arrows with a code name or unit designation. | ___ | ___ |
| **8.** Labeled the MSR and assigned a code name. | ___ | ___ |
| **9.** Ensured that the center of a unit's symbol corresponded to the coordinates of the unit. | ___ | ___ |

**Evaluation Guidance:** Refer to chapter 1, paragraph 1-4 b (6).

**References**

**Required:** FM 1-02 and FM 3-90.5 (FM 3-90.2)

**Related:**

**SUBJECT AREA 20: DEFENSE MEASURES**

## 071-430-0002
### Conduct a Defense by a Squad

**Conditions:** Given a squad, a priority of work, and locations for crew-served weapons designated by the platoon leader.

**Standards:** Accomplish preparation of a defensive position within the time specified in the platoon leader's order while maintaining security, camouflage, and concealment.

1. Designate fighting positions for squad members.
2. Designate alternate and supplementary positions for squad members.
3. Ensure assigned priority of work is followed by all squad members.
4. Maintain security.
5. Continue work as rapidly as possible.
6. Maintain camouflage and concealment (to include noise, light, and litter discipline).
7. Construct positions properly.

**Performance Steps**

**1.** Designate primary positions.

**a.** Ensure that each position has cover, concealment, and good fields of fire. Position weapons so fires overlap, mutually support one another, and can be integrated to place heavy surprise fire on the enemy. Watch closely any routes that could provide the attacker cover and positions from where the attacker can mass fire on your position. Be sure your entire squad sector is covered so you can repel any assault.

**b.** Consider the number of Soldiers available, and then position them so that each Soldier supports the Soldier to their right and left. When the platoon's key weapons have been positioned, the Soldiers in the squad are positioned to protect those weapons from a dismounted assault. Each sector of fire must cross in front of another position at a point beyond hand grenade range. Site each fighting position using natural cover and concealment; then—

**(1)** Clear fields of fire to allow each Soldier to do the job for which they are positioned. DO NOT OVER CLEAR.

**(2)** Build artificial cover such as a parapet, if required.

**(3)** Hide everything.

**c.** Ensure that all enemy approaches into your squad sector are adequately covered by automatic weapons fire.

**d.** Before digging in, move to the front of the position to ensure that each fighting position has frontal cover from enemy fire and that all positions are mutually supporting.

***Note***: A good fighting position should allow you to see and fire to the front when not receiving effective direct fire. However, if effective direct fire is received, the Soldier can move behind frontal cover and fire to the oblique.

**e.** To position each weapon of a rifle squad, follow the procedures below.

**(1)** Automatic riflemen. Identify dismounted avenues of approach such as ravines, draws, and heavily wooded or brushy areas that are not covered with the machine gun. If possible, find out what dead space the machine gun has and cover it with automatic rifle fire. If there is no dead space, assign a sector that will interlock with and overlap the machine gun's sector and final protective fire.

**(2)** Grenadiers. Position grenade launchers to cover the dead space of the automatic weapon's final protective fire. They must also be positioned to cover the entire squad sector.

**(3)** Riflemen. Position riflemen between the remaining positions to give continuous observation and fire throughout the squad sector. They provide mutual support between positions and identify targets for the squad and platoon's key weapons (grenade launcher, squad automatic weapons, machine gun, and Dragons).

**(4)** Claymore mines. Use these to cover any dead space that cannot be covered with 40-mm grenade launchers and to supplement the fire of your riflemen.

**Performance Steps**

**f.** Select a position where you can control the fire of your squad. Use your weapon only when necessary for protection, to direct fire, or to influence the action at a critical point. Position yourself slightly behind the squad so you can observe and contact your squad, or at least your team leaders. If your squad's strength is reduced, or the terrain does not permit you to establish a position to the rear, you must man a forward position. In this case, you may have to construct a single position to be able to observe your squad sector. You must be able to maintain contact with your platoon and fire team leaders from whatever position you select. Look for covered routes that you can use to move to the positions of your team leaders and platoon leader.

**2.** Designate alternate positions.

**a.** Prepare alternate positions so they are ready for occupation by the squad when the primary positions are no longer defendable. The locations of alternate positions highly depend on the terrain, cover and concealment, and the existing enemy situation. They must be near enough to the primary positions so the squad can cover the same sectors of fire as from the primary positions without sustaining excessive casualties. A Soldier's alternate position may be to the flank or slightly to the rear of their primary position.

**b.** When selecting alternate positions, consider the following points:

**(1)** Do they allow the weapon(s) or element(s) to accomplish the same mission as from the primary positions?

**(2)** Do they provide—

**(a)** Observation of the primary sector of fire?

**(b)** Cover and concealment?

**(c)** Maximum use of natural and artificial obstacles?

**(d)** Control of the key terrain in the squad sector?

**(e)** Coverage of the avenues of approach into the sector?

**(f)** Cover and concealment of withdrawal?

**c.** Designate squad alternate positions based on the alternate positions of key weapons (M60 machine gun, M249 machine gun, and Javelin).

**d.** Depending on their priority, prepare alternate positions immediately after completing the primary positions. Communication trenches should be constructed between primary and alternate positions as time and terrain permit. When supervising the preparation of alternate positions, ensure that they are sited and constructed to take maximum advantage of the natural defensive characteristics of the terrain and the capabilities of organic weapons.

| Performance Steps |
|---|

**3.** Designate supplementary positions.

**a.** Orient supplementary positions (unlike alternate positions) in a different direction than the primary position. Supplementary positions will normally be within 200 meters of primary positions. As time and terrain permit, prepare communication trenches to provide covered routes between primary and supplementary positions.

**b.** When selecting supplementary positions, consider the following:

**(1)** Do they allow the squad to defend as part of the platoon against enemy attack from the flank(s) and rear?

**(2)** Do they cover the most dangerous avenues of approach, other than those into the primary positions?

**(3)** Do they meet the same guidelines that apply to placement and construction of primary and alternate positions?

**Evaluation Preparation:** *Setup*: Select an area in the field with varying terrain, cover and concealment. Show the squad leader the squad's area of responsibility.

*Brief Soldier:* Tell the Soldier to select primary, alternate, and supplementary positions, in an area specified by the platoon leader, ensuring that the supplementary position is oriented in a different direction than the primary position.

| Performance Measures | GO | NO GO |
|---|---|---|
| **1.** Ensured that the primary position met the following requirements: | —— | —— |
| **a.** Used natural cover and concealment. | | |
| **b.** Ensured that all fighting positions and weapons emplacements were mutually supporting. | | |
| **c.** Marked and informed squad members where fighting positions were constructed. | | |
| **d.** Supervised construction of the fighting positions. | | |
| **2.** Ensured that the alternate position met the following requirements: | —— | —— |
| **a.** Provided observation of the primary sector of fire. | | |
| **b.** Provided cover and concealment. | | |
| **c.** Provided use of natural and artificial obstacles. | | |
| **d.** Provided coverage of avenues of approach into the sector. | | |
| **e.** Provided covered and concealed routes of withdrawal. | | |

| Performance Measures | GO | NO GO |
|---|---|---|
| **3.** Ensured that the supplementary position provided for and met the following requirements:<br><br>**a.** Allowed the squad to defend as part of the platoon against enemy attack from the flank(s) and rear.<br><br>**b.** Provided observation of the primary sector of fire.<br><br>**c.** Provided cover and concealment.<br><br>**d.** Provided natural and artificial obstacles.<br><br>**e.** Provided coverage of avenues of approach into the sector.<br><br>**f.** Provided covered and concealed routes for withdrawal.<br><br>**g.** Covered dangerous avenues of approach other than those into the primary position.<br><br>**h.** Provided the same guidelines in construction and placement of the supplementary position as in the alternate position. | ___ | ___ |

**Evaluation Guidance:** Refer to chapter 1, paragraph 1-4 b (6).

**References**

**Required:** FM 7-7 and FM 3-21.8 (FM 7-8)

**Related:**

**SUBJECT AREA 22: UNIT OPERATIONS**

---

## 071-326-5805

### Conduct a Route Reconnaissance Mission

**Conditions:** Given a platoon, a 1:50,000 map of the area of operation, and a mission to conduct a route reconnaissance.

**Standards:** Plan and conduct a route reconnaissance well enough to—

1. Organize the platoon to conduct the reconnaissance mission.
2. Use movement techniques appropriate for the likelihood of enemy contact.
3. Obtain necessary information concerning the conditions, obstacles, critical terrain features, and enemy along the assigned route.

**Performance Steps**

**1.** Plan the reconnaissance. Receive the order.

**a.** Issue a warning order

**b.** Gather information and prepare an operation order based on the factors of mission, enemy, terrain and weather, troops and support available, time available, civil considerations (METT-TC).

**c.** Ensure that the plan is as detailed as possible and includes the exact information to obtain, the time to report the information, where to report the information, where to seek the information action to take on enemy contact, and when to execute the mission. Essential details include—

**(1)** Pertinent information on the enemy, friendly troops, and the area of operations.

**(2)** Proposed plans of higher commands, to include anticipated traffic flow along the route and types of vehicles to be employed.

**(3)** When, where, and how to report information.

**(4)** Time of departure.

**(5)** Appropriate control measures.

**(6)** Action to take after completing the mission.

**(7)** Special equipment requirements.

**(8)** Terrain considerations of existing routes and their physical characteristics:

**(a)** Gradients of slope and the radius of curvature.

**(b)** Bridges.

**(c)** Vehicular fording, ferrying, and swimming sites.

**(d)** Tunnels, under passes, and similar obstructions to traffic flow.

**(e)** Artificial obstacles (such as areas of chemical, biological, and radiological contamination, roadblocks, craters, and minefields).

**(f)** Rock falls and slide areas.

**(g)** Drainage.

**(h)** Other natural or manmade features (such as wooded and built-up areas) that may affect movement.

**2.** Supervise the preparation of the route reconnaissance. Ensure that Soldiers—

**a.** Maintain weapons and equipment on vehicles.

**b.** Conduct inspections.

**c.** Conduct rehearsals.

**Performance Steps**

**d.** Issue the order.

**e.** Ensure that the rest plan is followed.

**3.** Control the route reconnaissance.

**a.** Ensure that—

**(1)** Reconnaissance commences from the flanks or rear when reconnoitering areas along the route are not likely to be defended by enemy detachments (such as bridge approaches, defiles, or built-up areas).

**(2)** Detailed observation precedes actual reconnaissance.

**(3)** Approach routes are checked for mines, booby traps, and signs of ambush.

**b.** Ensure that when time is available, dismounted personnel are sent forward first, covered by the remaining elements of the unit. The number of dismounted personnel sent forward depends on the size of the objective and on the available approaches, cover, and concealment. If the dismounted patrols find that the near edge of the area is clear, the remainder of the unit moves quickly forward. The dismounted patrols then continue the reconnaissance and overwatch. The remainder of the unit closely follows the dismounted patrols.

**c.** When conducting a mounted reconnaissance, part of the unit remains mounted and moves forward cautiously but rapidly, overwatched by the remaining mounted elements. If the near edge of the area is clear, the overwatching elements moved forward quickly and the advance continues.

**4.** When conducting reconnaissance by fire, follow these procedures:

**a.** Reconnaissance by fire is accomplished by firing on likely or suspected enemy positions in an attempt to remove camouflage and to cause the enemy to disclose his/her presence by movement or return fire. During reconnaissance by fire, positions being reconnoitered are observed continuously so that enemy activity can be quickly and definitely located.

**b.** Reconnaissance by fire is employed by route reconnaissance teams as a security measure when time is critical and the loss of surprise is not essential.

**c.** If the enemy returns fire, the situation is further developed. If the fire is not returned, reconnaissance continues. Caution is exercised to ensure that the presence of a well-disciplined enemy is not overlooked.

**Evaluation Preparation:** *Setup*: At the test site, provide all personnel, equipment, and material given in the task condition statement.

*Brief Soldier:* Tell the Soldier to plan and conduct a route reconnaissance using proper movement techniques for the likelihood of enemy contact and to gather information about enemy forces along the assigned route and critical terrain features.

| Performance Measures | GO | NO GO |
|---|---|---|
| **1.** Planned reconnaissance based on the intelligence requirement. | ___ | ___ |
| **2.** Issued the plan for the mission. | ___ | ___ |
| **3.** Conducted hasty or deliberate reconnaissance based on time available and detail requirements. | ___ | ___ |
| **4.** Used reconnaissance by fire when permitted by the tactical situation. | ___ | ___ |
| **5.** Avoided decisive engagement with enemy forces. | ___ | ___ |
| **6.** Used proper movement techniques during the route reconnaissance. | ___ | ___ |
| **7.** Reported all items of military significance. | ___ | ___ |

**Evaluation Guidance:** Refer to chapter 1, paragraph 1-4 b (6).

**References**

**Required:** FM 3-21.20 (FM 7-20), FM 3-21.71, FM 3-90.1, FM 3-34.170 (FM 5-170), FM 7-7, FM 3-21.8 (FM 7-8), and FM 17-95

**Related:**

---

## 071-332-5021

### Prepare a Situation Map

**Conditions:** In a combat environment, given the tactical situation.

**Standards:** Include the following elements on the situation map:

1. All military symbols according to the current FM.
2. The enemy situation depicted in red or with double lines.
3. Control measures.
4. Location of the command post and/or command group.
5. Location of all maneuver elements.
6. Location of reserves.

| Performance Steps |
|---|
| **1.** Depict the control measures. |
| **2.** Depict the command post and/or the command group location. |
| **3.** Depict the location of the maneuver elements. |
| **4.** Depict the location of the reserve. |
| **5.** Determine the classification. |

**Evaluation Preparation:** *Setup*: At the test site, provide a copy of the operation order, map board, map, marking pens, overlay paper, coordinate scale, unit standing operating procedure (SOP), and FM 1-02 (FM 101-5-1).

*Brief Soldier:* Tell the Soldier to prepare a situation map. The situation map must be prepared according to FM 1-02 (FM 101-5-1) and the unit SOP.

| Performance Measures | GO | NO GO |
|---|---|---|
| **1.** Depicted the control measures. | —— | —— |
| **2.** Depicted command post and/or command group location. | —— | —— |
| **3.** Depicted the location of the maneuver elements. | —— | —— |
| **4.** Depicted the location of the reserve. | —— | —— |
| **5.** Determined classification. | —— | —— |

**Evaluation Guidance:** Refer to chapter 1, paragraph 1-4 b (6).

**References**

**Required:**

**Related:** FM 1-02 (FM 101-5-1) and FM 5-0

## 551-88M-3601

**Perform Duties as Convoy Commander**

**Conditions:** In a contemporary operating environment (COE), given a mission by the company commander to serve as a convoy commander; vehicles and operators; subordinate noncommissioned officers (NCOs) to serve as serial/march unit commanders; information from the commander's operation order (OPORD); map overlays; cargo to transport; and a timeline to follow.

**Standards:** Plan, prepare and execute the mission of commanding a convoy in COE: 1) notify all supporting elements to determine ability to support mission, 2) clearly define intent and requirements of mission operations order, 3) ensure that sufficient number of vehicles is allocated for mission, 4) ensure that a route reconnaissace is performed by designated personnel, 5) brief all convoy participants on their duties and responsibilities, 6) resolve all issues that may hinder or prevent successful completion of mission. Complete all actions while adhering to convoy timeline without injury to personnel or damage to equipment.

| Performance Steps |
|---|

**1.** Plan the convoy movement.

**a.** Consider the following as key factors in the planning process of a convoy.

**(1)** Mission, enemy, terrain and weather, troops and support available, time available, civil considerations (METT-TC).

**(2)** The status of the training of drivers.

**(3)** Types of loads.

**(4)** Number of vehicles involved.

**(5)** Traffic conditions.

**(6)** Quality of road networks.

**Performance Steps**

**(7)** Advanced/quartering party.

**(8)** Convoy control personnel.

**(9)** Start and release points.

**(10)** Halts.

**(11)** Gaps and march rate.

**(12)** Submission of movement bid (if applicable).

**(13)** Communications.

**(14)** Route reconnaissance.

**(15)** Escort and security elements.

**(16)** Convoy support.

***Note***: The convoy commander should always refer to the unit standing operating procedures (SOPs) for standardized information concerning convoy guidelines. Whenever the SOP fails to provide the needed information, the convoy commander should solicit information from the unit chain of command and applicable regulations.

**b.** Review the OPORD.

**(1)** Determine the number of vehicles needed.

**(2)** Determine the number of personnel required.

**(3)** Determine if a special hauling permit is required.

**(4)** Determine the necessity of available supporting elements (fire support, close air support, engineer, chemical, or refueling support).

**(5)** Determine if a movement bid or convoy clearance is necessary. If so, submit request (DD Form 1265 [*Request for Convoy Clearance*]). This requirement is based on local guidelines (SOP) and the selected route for the convoy. In North Atlantic Treaty Organization (NATO) controlled areas, standardization agreement (STANAG) 2154 and STANAG 2155 are applicable. Submit at least 10 days prior to planned move.

**(6)** Determine if a request for oversized vehicles/loads is required. If so, submit DD Form 1266 (*Request for Special Hauling Permit*). The request must reach the approving authority 15 days before the planned move.

**(7)** Determine the convoy route.

*c. Conduct a map and route reconnaissance of the convoy route. Use engineer reconnaissance report information (DA Form 1711 [Engineer Reconnaissance Report]), if available, to determine route characteristics that may play a key part in your convoy mission.*

**(1)** Select an alternate route, if possible.

**(2)** Determine all critical points/checkpoints.

**(3)** Determine situational requirements.

**(4)** Determine choke points along the route that may endanger the mission.

**(5)** Plot all necessary items on a map.

**(6)** Plot scheduled halts.

**d.** Perform a risk management according to FM 5-19.

**Performance Steps**

**e.** Determine the size of serial/march units.

**f.** Determine the march rate (if not covered by the SOP).

**g.** Determine the vehicle march order, interval, and location of key vehicles within the convoy.

**h.** Determine the predeparture assembly area. Seek command guidance if information is not provided in OPORD.

**i.** Determine security requirements. Consider the following:

**(1)** Noise and light discipline.
**(2)** Front, flank, and rear security.
**(3)** Security during halts.
**(4)** Air cover.
**(5)** Fire support.
**(6)** Communications security.
**(7)** Deception.

**j.** Determine necessity and availability of spare vehicle (bobtails) for recovery.

**k.** Determine maintenance support.

**l.** Determine key convoy chain of command duty positions. Include the following:

**(1)** Assistant convoy commander.
**(2)** Serial/march unit commanders.
**(3)** Pacesetter.
**(4)** Trail officer.
**(5)** Trail maintenance officer.

**m.** Delegate responsibility to construct strip maps for all drivers.

**n.** Determine if and how many route guides are necessary.

**o.** Determine time/distance factors. Consider driver limitations, maximum driving time per shift, and co-drivers.

**p.** Coordinate with squad leaders and maintenance sergeant and other platoon sergeants on availability of vehicles to support the movement.

**q.** Determine the preparation timeline for the following:

**(1)** Loading of vehicles.
**(2)** Marshaling vehicles.

**r.** Determine the makeup and duties of the advance/quartering party to be performed at the destination (as applicable). Consider the following:

**(1)** Noncommissioned officer in charge (NCOIC).
**(2)** Other key personnel.

**Performance Steps**

**(3)** Ensuring that the convoy is able to move quickly off the route and into the marshaling area.

**(4)** Positioning of vehicles within the marshaling area.

**2.** Prepare for convoy movement.

***Note***: The convoy commander must perform specific actions to prepare the convoy. A limited amount of time is available to accomplish the following: select and reconnoiter the route, submit a movement bid if required (DD Form 1265), effect coordination for en route security, give instructions to subordinate element commanders and other supervisory personnel, inspect personnel and vehicles, and brief convoy personnel).

**a.** Submit a request for convoy clearance and oversized vehicle/load permits (if applicable).

**b.** Assign key convoy duty positions within the convoy to include the following:

**(1)** Assistant convoy commander.

**(2)** Serial/march unit commanders.

**(3)** Pacesetter.

**(4)** Trail officer.

**(5)** Trail maintenance officer.

**(6)** Route guides (if necessary).

**(7)** Security NCOIC and force.

**c.** Brief and dispatch the advance/quartering party to destination (as applicable).

**d.** Ensure that the following predeparture actions are done prior to movement.

**(1)** Instruct the serial/march unit commander(s) to do the following:

**(a)** Supervise and spot check all preventive maintenance of convoy vehicles.

**(b)** Ensure that all vehicles are dispatched according to the unit SOP.

**(c)** Ensure that all personnel involved in the convoy are prepared and have in their possession all items identified by unit SOP for the convoy mission.

**(d)** Ensure that all vehicles, to include trailers and support vehicles, are loaded and prepared for movement as directed.

**(e)** Ensure that all personnel have in their possession all applicable accident and load forms required by the unit SOP and regulation (DD Form 518 [*Accident Identification Card*] and SF Form 91 (*Motor Vehicle Accident Report*) and DD Form 626 [*Motor Vehicle Inspection*] and DD Form 836 [*Dangerous Goods Shipping Paper/Declaration and Emergency Response*] [if transporting hazardous material [HAZMAT] prior to movement).

**Performance Steps**

**(f)** Ensure that all vehicles are marked with convoy clearance numbers (if movement is over a controlled route).

**(g)** Ensure that convoy control vehicles are marked with the appropriate flags, signs, or lights (as applicable).

**(h)** Relay all necessary communications information.

**(i)** Hand out all strip maps to vehicle crews.

**(j)** Perform sensitive items check (to be verified during convoy briefing).

**(2)** Once all vehicles, loads, and personnel have been checked, supervise the staging of vehicles in the designated assembly area by order of march and serial/march unit.

**(3)** Coordinate to obtain artillery support (if available).

**e.** Ensure that all of the following questions are answered (Convoy Commander's Checklist) prior to movement.

**(1)** Where is the start point (SP)? The release point (RP)?

**(2)** What route is to be used?

**(3)** Has reconnaissance been made? Has the condition of the route been determined?

**(4)** Can bridges, tunnels, underpasses, and defiles safely accommodate all loaded and tracked (if applicable) vehicles?

**(5)** Are critical points known and listed on the map?

**(6)** What is the size of serials?

**(7)** What is the size of march units?

**(8)** What is the rate of march?

**(9)** What is the vehicle interval on an open road? Built-up road?

**(10)** What type of column will be used?

**(11)** What provision has been made for refueling (if applicable)?

**(12)** Has a suitable operations area been selected?

**(13)** Have suitable rest and mess-halt area been selected?

**(14)** Have convoy clearances been obtained? What date?

**(15)** Is an escort required? Has it been requested?

**(16)** Are spare trucks available for emergencies?

**(17)** Are vehicles fully serviced and ready for loading?

**(18)** Are loads properly blocked and braced, neat, and balanced?

**(19)** Are drivers properly briefed? By whom? When? Are strip maps furnished?

**(20)** Is the convoy marked front and rear of each march unit? With convoy number when required? Is each vehicle marked? Are convoy flags on the vehicles?

**(21)** Are guides in place? Have arrangements been made to post and recover them?

**(22)** Are blackout lights functioning?

**(23)** Have maintenance services been alerted?

**(24)** Is the maintenance truck in the rear? Are medics in the rear?

**(25)** Are all interested parties advised of the estimated time of arrival (ETA)?

**(26)** Is the officer at the rear of the convoy ready to take necessary corrective action (such as investigating accidents and unusual incidents and changing loads)?

**(27)** Who is the trail officer?

**(28)** Is there a truck unloading plan? Who is responsible? Do they have the necessary equipment?

**(29)** Is there a plan for feeding personnel?

**(30)** Have times been established for loading trucks?

**(31)** Has time been established for formation of convoy?

**(32)** Have times been established for unloading trucks?

**(33)** Has time been established for releasing trucks? Who is responsible?

**(34)** Is there a carefully conceived plan known to all convoy personnel that can be used in case of an attack?

**(35)** Is a written OPORD, if required, on hand?

**(36)** Will a log of road movement be required at the end of the trip? Are necessary forms on hand?

**(37)** Has a weather forecast been obtained?

**(38)** Do all personnel have proper clothing and equipment?

**(39)** Is there a communications plan? Where will communications equipment be located? Has all communications equipment been serviced?

**(40)** If transporting hazardous materials, have all involved vehicles been appropriately placarded? Are load manifests in the driver's door pocket or otherwise readily available and known by convoy personnel?

**(41)** Have vehicles containing HAZMAT been placed at appropriate locations within the convoy to reduce residual damage in case of enemy attack?

**(42)** Has every effort been made to camouflage vehicles to the extent possible during movement? At halts?

**f.** Conduct convoy briefing.

***Note***: The convoy commander's briefing is given after all other movement preparations have been completed and verified by all responsible parties and the convoy is prepared to move.

***Note***: Gather applicable information from the commander's OPORD and local SOP to complete your convoy briefing. Provide as much applicable information as possible that may affect the convoy movement. Apply the information in the following format.

**(1)** Situation.

**(a)** Enemy forces.

**(b)** Friendly forces.

**(c)** Support units.

**Performance Steps**

**(2)** Mission.

**(a)** Type of cargo (to include hazardous materials precautions and guidelines).

**(b)** Origin.

**(c)** Destination.

**(3)** Execution.

**(a)** General organization.

**(b)** Time schedule.

**(c)** Convoy speed.

**(d)** Catch-up speed.

**(e)** Vehicle distance.

**(f)** Emergency measures (for accidents, breakdowns, and separation from the convoy).

**(g)** Actions of convoy and security personnel if attacked.

**(h)** Medical support.

**(4)** Administration and logistics.

**(a)** Personnel control.

**(b)** Billeting.

**(c)** Messing.

**(d)** Refueling and servicing of vehicles, complying with spill prevention guidelines.

**(5)** Command and signal.

**(a)** Convoy commander's location.

**(b)** Assistant convoy commander's designation (succession of command).

**(c)** Actions of security force commander.

**(d)** Serial commander's responsibilities.

**(e)** Arm and hand signals.

**(f)** Other prearranged signals.

**(g)** Radio frequencies and call signs (for control personnel, security force commanders, fire support elements, reserve security elements, and medical evacuation).

**(6)** Safety.

**(a)** Hazards of the route.

**(b)** Weather conditions.

**(c)** Defensive driving.

**Performance Steps**

(7) Environmental protection.

(a) Spill prevention.

(b) Transporting HAZMAT.

3. Conduct the convoy movement.

*Note*: Convoy commander should maintain a log of events during the convoy that may be required in the convoy commander's report upon completing the convoy movement.

*Note*: The convoy commander must be able to monitor and control all aspects of vehicle operations within the convoy. The convoy commander is ultimately responsible for ensuring that all assets arrive safely at the destination with a minimal amount of losses. The convoy commander's ability to delegate authority and enforce march discipline are key factors in mission accomplishment.

**a.** Conduct a communications check of all systems in the convoy radio net. Correct all communications deficiencies on the spot.

**b.** Signal all drivers to start engines.

**c.** Give the signal to begin movement and depart the assembly area (at the time designated in the movement order). Use the closed column formation until entry onto the main convoy route. If expressways are used, instruct drivers to close to approximately a 20-meter vehicle interval when entering the acceleration ramp.

**d.** Ensure that the convoy reaches the SP according to the established timeline.

**e.** Monitor radio traffic.

**f.** Ensure that the trail officer relays passing the SP on the established timeline.

**g.** Ensure that the pacesetter maintains the established speed.

**h.** Signal drivers to adjust speed and interval accordingly.

**i.** Notify higher headquarters upon passing each checkpoint/critical point on the route (as directed by the SOP or commander).

**j.** During halts, ensure that serial/march unit commanders (if entire convoy is at halt) complete halt checks on personnel, vehicles, and loads.

**(1)** Ensure that vehicles are staged so as to facilitate rapid movement. Vehicles should remain in the same order of march as during movement.

**(2)** Ensure that security is posted to prevent pilferage or compromise by enemy forces.

**(3)** Exchange drivers if operating limits have been reached. Do not exceed authorized driving times as listed in AR 385-10, paragraph 11-4.

**Performance Steps**

**k.** As terrain dictates, direct the column into open or closed formations. Movement through urban areas will facilitate closing the formation to established vehicle intervals.

**l.** Enforce passive and active defense measures within the convoy.

**m.** Enforce operations security (OPSEC) and communications security (COMSEC).

**n.** Report any enemy contact made during movement to higher headquarters. Include attack location, size of the enemy force if known, types and number of weapons used, damage inflicted on convoy assets, reactive measures taken, casualties incurred, and any information covered in the unit SOP.

**o.** As the convoy passes the RP, ensure that vehicle accountability is conducted and that control is relinquished (as applicable) for those elements (supply convoy) in which custody and control will change and elements may continue to other destinations.

**p.** Contact trail/trail maintenance officer for updates on any/all vehicle breakdowns, actions taken, and status of effected equipment/loads/personnel.

**4.** Conduct convoy closure operations.

**a.** For those convoy assets that are relinquished upon crossing the RP, ensure that the chain of custody is not broken and that all command and control of released assets is delegated to authorized personnel.

**b.** For those convoy assets that remain under current convoy control upon crossing RP, the following steps apply.

**(1)** Ensure that personnel designated as ground guides (either assistant drivers or advanced party personnel) escort vehicles off the convoy route and into the assembly area in a timely manner to minimize congestion.

**(2)** Enforce security measures in the assembly area.

**(3)** Ensure that all vehicles are positioned in the assembly area as to facilitate security and for off-loading operations as necessary.

**(4)** Instruct serial/march unit commanders/squad leaders to conduct sensitive items checks and supervise after-operations maintenance on vehicles.

**(5)** Ensure that vehicles with hazardous cargo are positioned away from facilities according to regulations (DA Pam 385-64, paragraph 11-15).

**(6)** Maintain contact with trail/trail maintenance officer for closure through RP.

**(7)** Facilitate the recovery of damaged or other non-operational vehicles upon arrival at the assembly area.

**(8)** Coordinate with trail/trail maintenance officer for information regarding any vehicle accident reports that may be necessary (DA Form 285 [*U.S. Army Accident Report*]).

| Performance Steps |
|---|

**(9)** Report closure and convoy status to higher headquarters according to the unit SOP.

**(10)** If required, submit convoy commander's report to higher headquarters according to FM 55-15, chapter 3 and FM 55-30, appendix R.

**Evaluation Preparation:** *Setup:* Evaluate this task during a field training exercise or normal training session. Provide the Soldier with the items listed in the conditions statement.

*Brief Soldier:* Tell the Soldier he/she will be evaluated on his/her ability to properly perform duties as a convoy commander.

| Performance Measures | GO | NO GO |
|---|---|---|
| 1. Planned the convoy movement. | —— | —— |
| 2. Prepared for convoy movement. | —— | —— |
| 3. Conducted the convoy movement. | —— | —— |
| 4. Conducted convoy closure operations. | —— | —— |

**Evaluation Guidance:** Refer to chapter 1, paragraph 1-4 b (6).

**References**

**Required:** AR 55-162, DA Form 1711, DA Form 285, DA Pam 385-64, DD Form 518, DD Form 626, DD Form 836, DD Form 1265, DD Form 1266, FM 3-34.170, FM 3-100.12, FM 4-01.30, FM 5-0, FM 5-19, FM 21-60, FM 21-305, FM 55-15, FM 55-30, SF Form 91, STANAG 2155, and STANAG 2155

**Related:** AR 385-10

## 551-88N-3042
### Plan Unit Move

**Conditions:** You are given a command directive to plan for your unit to conduct a move to port of embarkation to deploy in support of an Army or Joint operations plan (OPLAN). You have access to the unit standing operating procedure (SOP), and all unit movement directives.

**Standards:** Plan a unit movement using the necessary references (vehicle -10s) and equipment to deliver vehicles and equipment at the port of debarkation (POD) with no loss of vehicles, equipment, or personnel.

| Performance Steps |
|---|

1. Identify what needs to be moved.
2. Identify the equipment to be moved.
3. Identify what needs to be moved by air.
4. Identify hazardous, sensitive, and classified equipment/material.
5. Identify bulk cargo.
6. Develop vehicle load plans.

| Performance Steps |
| --- |
| **7.** Translate what needs to be moved into transportation terms using Transportation Coordinators' Automated Command and Control Information System/ Transportation Coordinators' Automated Information for Movements System (TC ACCIS / TC-AIMS II). |
| **8.** Determine how the personnel and equipment will move to the aerial/seaport of debarkation (APOD/SPOD). |
| **9.** Prepare the unit movement plan. |
| **10.** Maintain the movement plan. |

**Evaluation Preparation:** *Setup*: Evaluate this task during a field training exercises or a unit training exercise.

*Brief Soldier:* Inform the Soldier that the evaluation is measured on the ability to perform all the functions listed.

| Performance Measures | GO | NO GO |
| --- | --- | --- |
| **1.** Identified what needed to be moved. | —— | —— |
| **a.** Personnel. | | |
| **b.** Equipment. | | |
| **c.** Supplies. | | |
| **2.** Identified the equipment to be moved: | —— | —— |
| **a.** What needs to accompany troops—Yellow (to accompany troops) TAT. | | |
| **b.** What is needed immediately upon arrival—Red TAT. | | |
| **c.** What does not have to accompany troops—NTAT. | | |
| ***Note:*** NTAT consists of all other equipment that is required for the unit to perform its mission. | | |
| **3.** Identified what needed to be moved by air. | —— | —— |
| **a.** Advance party personnel. | | |
| **b.** Main body personnel. | | |
| **c.** Baggage TAT. | | |
| **d.** Equipment. | | |

| Performance Measures | GO | NO GO |
|---|---|---|
| **4.** Identified hazardous, sensitive, and classified equipment and material. | ___ | ___ |
| **a.** Determined appropriate packaging, labeling, segregating, and placarding for movement. | | |
| **b.** Identified TAT ammunition quantities. | | |
| **c.** Identified vehicles (3/4 tank full sea/air). | | |
| **d.** Identified individual weapons (remain with the Soldier; bolt may be removed) | | |
| **e.** Identified crew-served weapons (palletized or carried in the baggage compartment). | | |
| ***Note***: Movement personnel must refer to applicable regulations to obtain the detailed information for planning and execution. | | |
| **5.** Identified bulk cargo. | ___ | ___ |
| **a.** Developed a packing list for all consolidated cargo loaded in vehicles, containers, and 463L pallets. | | |
| **b.** Determined packing list distribution. | | |
| **c.** Determined blocking, bracing, packing, crating and tiedown (BBPCT) requirements. | | |
| **6.** Developed vehicle load plans. | ___ | ___ |
| **a.** Planned for cross-country payload capacity. | | |
| **b.** Reduced vehicles according to the mode of transportation and type of movement. | | |
| **c.** Tested planned loads. | | |
| **d.** Weighed and recorded planned loads. | | |
| **e.** Identified transportation requirements exceeding the unit's organic lift capability. | | |
| **7.** Translated what needed to be moved into transportation terms using TC ACCIS/TC AIMS II: | ___ | ___ |
| **a.** Used the TC ACCIS to translate the automated unit equipment list/deployment equipment list (AUEL/DEL). | | |
| **b.** Used the TC AIMS II to translate the organization equipment list (OEL)/unit deployment list (UDL). | | |

| Performance Measures | GO | NO GO |
|---|---|---|
| **8.** Determined how the personnel and equipment will move to the APOD/SPOD). | —— | —— |
| **a.** Roadable vehicles. | | |
| **b.** Tracked vehicles. | | |
| **c.** Rotary wing aircraft. | | |
| **9.** Prepared the unit movement plan. Determined administrative, logistical and coordinating requirements for the plan such as transportation for drivers from the APOD/SPOE and back to the unit and for petroleum, oil and lubricants (POL). | —— | —— |
| **10.** Maintained the movement plan. Kept the AUEL/OEL current with changes in unit equipment, personnel, and supplies. | —— | —— |

**Evaluation Guidance:** Refer to chapter 1, paragraph 1-4 b (6).

**References**

**Required:** FM 4-01.011, FM 5-0, FM 55-1, FM 100-17, FORSCOM Reg 55-2, and TB 55-46-1

**Related:** FORSCOM Reg 55-1

# 805C-PAD-3594

## Store Classified Information and Materials

**Conditions:** You are a squad/section leader. Given classified material, simulated storage containers, and AR 380-5.

**Standards:** Secure classified information and materials in appropriate storage containers.

**Performance Steps**

1. Determine classification levels of information and materials to be stored.
2. Determine the amounts of classified information and materials to be stored.
3. Determine the method of storage/safeguarding to be used.
4. Determine the type of facility or storage containers required.
5. Determine the appropriate lock(s) for storage facility or containers.
6. Determine initial or additional facility/equipment requirements.

| Performance Steps |
|---|

**7.** Submit a request for needed facility/equipment through the appropriate channel.

**a.** Determine routing.

**b.** Prepare the request in the proper format.

**c.** Submit the request in distribution or as required by local SOP.

**8.** Restrict access to facility/materials.

**a.** Determine personnel authorized access to storage facility/containers.

**b.** Verify the level of classified information and/or materials personnel are authorized access.

**9.** Maintain physical security.

**a.** Secure all classified material when not in use.

**b.** Indicate when containers are locked or open.

**c.** Designate personnel responsible for securing and controlling classified facility/containers.

**10.** Maintain records of security checks of classified facility/containers.

**a.** Place the security check form on the facility and containers as required.

**b.** Designate personnel to conduct daily/weekly checks.

**c.** Verify that security checks are made.

**11.** Train section personnel on security requirements for classified materials.

**a.** Provide requirements/authorization for access to facility/materials.

**b.** Provide instructions on conducting security checks.

---

**Evaluation Preparation:** *Setup*: To evaluate this task give a scenario that would require the Soldier to perform the performance measures. This will require simulated containers, locks, DA Form 702 (*Security Container Check Sheet*), a list of section personnel with varying security clearances, documents marked with various levels of security classification (marked "classified for training purposes only"), format for security check assignment, and classified document covers.

*Brief Soldier:* Tell the Soldiers that he/she will be evaluated on his/her ability to select the appropriate container for each document concerned and indicate who would have access to which container.

| Performance Measures | GO | NO GO |
|---|---|---|
| **1.** Determined classification levels of information and materials to be stored | ____ | ____ |
| **2.** Determined the amounts of classified information and materials to be stored. | ____ | ____ |

| Performance Measures | GO | NO GO |
|---|---|---|
| **3.** Determined the method of storage/safeguarding to be used. | ___ | ___ |
| **4.** Determined the type of facility or storage containers required. | ___ | ___ |
| **5.** Determined appropriate lock(s) for storage facility or containers. | ___ | ___ |
| **6.** Determined initial or additional facility/equipment requirements. | ___ | ___ |
| **7.** Submitted a request for needed facility/equipment through the appropriate channel. | ___ | ___ |
| **a.** Determined routing. | | |
| **b.** Prepared the request in the proper format. | | |
| **c.** Submitted the request in distribution or as required by the local SOP. | | |
| **8.** Restricted access to facility/materials. | ___ | ___ |
| **a.** Determined personnel authorized access to storage facility/containers. | | |
| **b.** Verified the level of classified information and/or materials personnel are authorized access. | | |
| **9.** Maintained physical security. | ___ | ___ |
| **a.** Secured all classified material when not in use. | | |
| **b.** Indicated when containers were locked or open. | | |
| **c.** Designated personnel responsible for securing and controlling classified facility/containers. | | |
| **10.** Maintained records of security checks of classified facility/containers. | ___ | ___ |
| **a.** Placed the security check form on the facility and containers as required. | | |
| **b.** Designated personnel to conduct daily/weekly checks. | | |
| **c.** Verified that security checks were made. | | |

| Performance Measures | GO | NO GO |
|---|---|---|
| **11.** Trained section personnel on security requirements for classified materials. | ___ | ___ |
| **a.** Provided requirements/authorization for access to facility/materials. | | |
| **b.** Provided instructions on conducting security checks. | | |

**Evaluation Guidance:** Refer to chapter 1, paragraph 1-4 b (6).

**References**

**Required:** AR 380-5

**Related:**

**SUBJECT AREA 23: SECURITY AND CONTROL**

## 191-379-4407

**Plan Convoy Security Operations**

**Conditions:** You are given an operation order (OPORD), a combat load, a map of the area, a lensatic compass, a protractor, communications equipment, signal operating instructions (SOIs), security personnel, and special orders, if required.

**Standards:** Develop a plan that includes all required considerations for the specific mission, coordinate with appropriate units for required support, ensure that necessary equipment and supplies are available and operational, and prepare briefings for the security personnel on the mission and assign specific duties. Ensure that 360-degree security is maintained from the point of origin to the destination when escorting special weapons.

| Performance Steps |
|---|

**1.** Receive the OPORD.

*Note*: Follow all steps in the troop-leading procedures. When carrying special weapons or ammunition, ensure that you obtain the special orders concerning this cargo, that you understand them, and that they cover any situation not covered in the OPORD.

**2.** Coordinate with host nation (HN) security personnel.

**3.** Consult with all sources of information, especially the engineers and the highway traffic division (HTD), to obtain as much information as possible.

**4.** Reconnoiter the convoy route.

**a.** Identify likely trouble spots and ambush sites.

**b.** Determine possible locations for traffic control posts (TCPs) and/or checkpoints.

**c.** Identify route conditions.

**d.** Determine the location of friendly units in the area.

**Performance Steps**

**5.** Coordinate with the convoy commander.

**a.** Determine the actions to take if—

**(1)** Attacked by a sniper.

**(2)** Ambushed with the road blocked.

**(3)** Ambushed with the road not blocked.

**(4)** Attacked from the air.

**(5)** Attacked with artillery.

***Note***: Based on the mission and/or the type of cargo, the reaction to an enemy attack may vary. When carrying special weapons or ammunition, ensure that procedures are covered in the special orders on exactly how to react to enemy contact.

**b.** Determine the protective measures to take for mines and booby traps.

**c.** Determine convoy organizations, to include the location of—

**(1)** Critical cargo vehicles.

**(2)** Control vehicles.

**(3)** Armored vehicles and automatic weapons.

***Note***: Armored vehicles and automatic weapons are positioned within the convoy so that they are mutually supporting.

**(4)** Maintenance and recovery vehicles.

***Note***: The convoy commander and/or unit maintenance officer or noncommissioned officer will decide whether to repair, recover, or destroy damaged vehicles.

**d.** Determine the primary and backup frequencies.

**e.** Determine emergency communications procedures.

**f.** Determine the timetable for movements, especially for serials and march units.

**g.** Identify the coordination points (assembly area for everyone).

**h.** Determine the start and release points.

**i.** Determine the security measures to be used at halts and rest stops. The location of the halts should be—

**(1)** In a relatively secure area.

**(2)** Off the roadway where there is some natural cover and concealment.

**(3)** Under the surveillance of a security force.

**(4)** In an area where there is a view for 200 meters at each end of the convoy with no obstructions, such as curves and hills.

**(5)** In an area that is large enough to maintain the proper convoy interval.

**(6)** In an area where local civilians can be kept away from the convoy.

**Performance Steps**

(7) In an area that is not heavily populated.

(8) In an area where HN police are present, if applicable.

*Note*: Additional procedures for establishing an exclusion area and enforcing the "two-man rule" must be covered and written in the special orders to ensure that 360-degree security is maintained when carrying special weapons or ammunition. If security cannot be maintained, instructions on how and when to disable the weapons must be included.

**j.** Determine the time and place that military police support begins and ends.

**k.** Determine any expected changes in the routes.

**l.** Determine road conditions.

**m.** Determine supporting fires, to include artillery support, engineer support for minesweeping of the route, gunship support, and HN police support, if applicable.

**n.** Identify the primary and alternate routes.

**o.** Determine the location where military police vehicles and personnel will be positioned in the convoy.

**p.** Determine the method of escort for the convoy. Escort methods are as follows:

(1) Scout, lead, and trail (leading/following).

(2) Empty truck (or modified).

(3) Leap frog.

(4) Perimeter.

**6.** Brief the personnel on the mission, enemy situation, and specific individual duties to be performed. Also brief personnel on—

**a.** Procedures to be followed in the case of mechanical breakdown.

**b.** Defensive measures to be used against mines and booby traps.

**c.** Actions to be taken when there is contact with the enemy.

**d.** The method of escort.

**e.** Communications.

**f.** Special orders, if applicable.

**7.** Coordinate with friendly units in the area where the convoy will pass.

**a.** Identify the support that friendly units can provide.

**b.** Identify the restrictions that apply to using indirect fire.

| Performance Steps |
|---|

**8.** Coordinate with artillery units.

**a.** Identify predetermined targets found during route reconnaissance or through coordination with the convoy commander.

**b.** Identify the locations for rest stops, halts, and possible enemy ambush sites.

**9.** Coordinate with helicopter and/or Air Force units for gunship support.

**10.** Inspect the following:

**a.** Vehicles.

**b.** Radios.

**c.** Ammunition.

**d.** Weapons.

**e.** Chemical, biological, radiological, and nuclear (CBRN) equipment.

**f.** Combat load.

**Evaluation Preparation:** *Setup*: Schedule a field training exercise that will require the squad leader to plan convoy security operations.

*Brief Soldier:* Tell the squad leader being evaluated that he/she must develop an accurate and concise plan for convoy security operations.

| Performance Measures | GO | NO GO |
|---|---|---|
| **1.** Received the OPORD. | —— | —— |
| *Note*: Followed all the steps in troop-leading procedures. | | |
| **2.** Coordinated with HN security personnel. | —— | —— |
| **3.** Consulted with all sources of information, especially the engineers and the HTD, to obtain as much information as possible. | —— | —— |
| **4.** Reconnoitered the convoy route. | —— | —— |
| **5.** Coordinated with the convoy commander. | —— | —— |
| *Note*: Ensured that 360-degree security was maintained from the point of origin to the destination when escorting special weapons. | | |
| **6.** Briefed personnel on the mission, enemy situation, and specific individual duties to be performed. | —— | —— |
| **7.** Coordinated with friendly units in the area where the convoy would pass. | —— | —— |
| **8.** Coordinated with artillery units. | —— | —— |

| Performance Measures | GO | NO GO |
|---|---|---|
| **9.** Coordinated with helicopter and/or Air Force units for gunship support. | ___ | ___ |
| **10.** Inspected equipment. | ___ | ___ |

**Evaluation Guidance:** Refer to chapter 1, paragraph 1-4 b (6).

**References**

**Required:**

**Related:** FM 3-19.1, FM 3-19.30, FM 3-19.4, FM 19-25, and FM 3-21.10 (FM 7-10)

## 301-371-1052

### Protect Classified Information and Material

**Conditions:** This task can be performed in field and garrison locations under all conditions. Given classified documents and/or material, AR 380-5, local standing operating procedures (SOPs), SF 700 (*Security Container Information*), SF 702 (*Security Container Check Sheet*), SF 703 (*Top Secret Cover Sheet*), SF 704 (*Secret Cover Sheet*), and SF 705 (*Confidential Cover Sheet*), DA Form 3964 (*Classified Documents Accountability Record*), office equipment and supplies.

**Standards:** Safeguard classified information/material—prevent unauthorized disclosure; maintain and store according to operation order (OPORD), AR 380-5, and the local unit standard operating procedures (SOPs).

| Performance Steps |
|---|

**1.** Identify classified material.

***Note***: Information is any knowledge that can be communicated or any documentary material (regardless of its physical form or characteristics) that is owned by, produced by or for, or is under the control of the United States Government. Unauthorized disclosure is a communication or physical transfer of classified information to an unauthorized recipient.

**a.** List the security classification levels.

**(1)** Confidential. Information, if disclosed to unauthorized persons, could reasonably be expected to cause damage to the national security.

**(2)** Secret. Information, if disclosed to unauthorized persons, could reasonably be expected to cause serious damage to the national security.

**(3)** Top Secret. Information, if disclosed to unauthorized persons, could reasonably be expected to cause exceptionally grave damage to the national security.

**b.** Define original and derivative classifications.

**(1)** Original classification—an initial determination that information requires, in the interest of national security, protection against unauthorized disclosure.

**(2)** Derivative classification—the incorporating, paraphrasing, restating, or generating in new form information that is already classified, and marking the newly developed material consistent with the classification markings that apply to the source information.

**Performance Steps**

**2.** Locate and read the declassification and downgrading instructions, which appear on a classified document. Define declassification and downgrading.

**a.** Declassification—the authorized change in the status of information from classified information to unclassified information.

***Note***: Information that already has been declassified and released to the public cannot be reclassified.

**(1)** At the time of original classification, the original classification authority shall attempt to establish a specific date or event for declassification.

**(2)** If a specific date or event for declassification cannot be determined, then information shall be marked for declassification 10 years from date of the original decision.

**(3)** The original classification authority may extend the duration of classification for a successive period, not to exceed 10 years at a time.

**(4)** At the time of original classification, the original classification authority may exempt for declassification within 10 years specific information that the unauthorized disclosure could reasonably be expected to cause damage to the national security.

**(5)** Information shall be declassified as soon as it no longer meets the standards for classification.

**b.** Downgrading—a determination by a declassification authority that information classified and safeguarded at a specific level shall be classified and safeguarded at a lower level.

**3.** Review information for required identification and markings.

***Note***: Classification markings will be in letters larger than those used in the rest of the text.

**a.** Check the document for required markings:

**(1)** Ensure that the overall classification of the document is marked, stamped, or affixed permanently—

**(a)** On the top and bottom of the outside front cover.

**(b)** On the title page.

**(c)** On the first page.

**(d)** On the outside of the back cover.

**(2)** Each interior page, except those left blank, will be marked on the top and bottom according to content, to include unclassified.

**(3)** Each section, part, paragraph, or similar portion of a classified document will be marked to show the level of classification of the information contained by showing the appropriate classification symbol.

**(4)** Charts, maps, and drawings will bear the appropriate classification marking for the legend, title, or scale block. The higher of these markings shall be inscribed at the top and bottom of each such document.

**Performance Steps**

**(5)** Photographs, films (including negatives), recordings, and their containers will be marked with the specific level of the information contained within.

**(6)** Information used to simulate classified or unclassified material or documents will be marked clearly to indicate the actual unclassified status of the information (for example, SECRET for Training Purposes Only).

**b.** Examine all classified documents for completeness; ensure that no parts or pages are missing.

**c.** Ensure that the following appears on the face of all classified information:

**(1)** The overall classification of the document should be stamped on the top and bottom.

**(2)** The identity, by name or personal identifier, and position of the original classification authority.

**(3)** The agency and office of origin.

**(4)** Declassification instructions.

**(5)** Reason for classification.

**4.** Protect classified information through restricted access.

**a.** Define access—the ability or opportunity to gain knowledge of classified information. Access is based on—

**(1)** Security clearance. Ensure that personal security clearance is equal to or exceeds the classification level of the material required to perform official duties.

**(2)** Need-to-know. Ensure that a determination has been made by an authorized holder of classified information that a prospective recipient requires access to specific information in order to perform or assist in a lawful and authorized government function.

***Note***: No one has a right to have access to classified information solely by virtue of rank or position.

**b.** Prevent unauthorized persons from gaining access to classified material.

**(1)** Keep all classified documents under constant observation when removed from storage.

**(2)** Keep all documents face down or covered when not in use.

**(3)** Place one of the following cover sheets on all classified material:

**(a)** SF 705 for confidential information.

**(b)** SF 704 for secret information.

**(c)** SF 703 for top secret information.

**c.** Define Special Access Program. This is a program for a specific class of classified information that imposes safeguarding and access requirements that exceed those normally required for information at the same classification level.

**Performance Steps**

**5.** Safeguard classified Information and material.

**a.** Use proper precautions to protect classified information and material:

**(1)** Safeguard classified information when temporarily departing the work area.

**(2)** Prevent display of classified information in public places.

**(3)** Use the following procedures when working papers containing classified information are created:

**(a)** Date the document.

**(b)** Mark the document with the words "Working papers."

**(c)** Mark each page, top and bottom, according to the content classification.

**(d)** Bring working papers under control as a finished document when—

- Retained more than 90 days from the date of origin.
- Released outside the originator's agency.
- Transmitted electrically or electronically.
- Filed permanently.
- Papers contain top secret information.

**6.** Comply with security regulations when discussing classified information.

**a.** Do not discuss classified and/or sensitive information on an unsecured telephone.

**b.** Use the following procedures when presenting a classified briefing:

**(1)** Ensure that the briefing area has been cleared to the highest level of material to be discussed.

**(2)** Check the attendance roster to ensure all personnel are cleared and have a need-to-know.

**(3)** Establish a sign-in roster at the main entrance to the briefing area and control access at all other entrances.

**(4)** Ensure that all training aids are marked with security classification according to their content.

**(5)** Inform the audience of the security classification of the briefing and the policy concerning note taking at the beginning of the briefing.

**(6)** Repeat the security classification at the end of the briefing.

**c.** Ensure that automated information systems (including networks and telecommunications systems) that collect, create, communicate, compute, disseminate, process, or store classified information have controls that—

**(1)** Prevent access by unauthorized persons.

**(2)** Ensure the integrity of the information.

**Performance Steps**

**7.** Protect classified information during transport.

**a.** Do not remove classified information from official premises without proper authorizations.

**b.** Protect classified information when transporting outside of a secured area.

**(1)** Address classified information to an official government activity or Department of Defense (DOD) contractor with a facility clearance and not an individual.

**(2)** Pack classified information in such a manner that the text will not be in contact with the inner envelope or container.

**(3)** Attach or enclose a receipt, DA Form 3964, in the inner envelope or container for all SECRET and TOP SECRET information. CONFIDENTIAL information requires a receipt only if the originator deems it necessary.

**(4)** Double wrap classified information in opaque envelopes or similar wrappings.

**(a)** Mark the inner envelope with the overall classification of the contents and any special instructions.

**(b)** Mark the inner and outer envelope with the complete sender's and receiver's address.

**(c)** Do not mark the outer envelope with any indications of the contents or the classification.

**(d)** Do not display classified material in public places while transporting.

**(e)** Do not store classified material in any detachable storage compartment (such as automobile trailers or luggage racks) while transporting.

**8.** Protect classified information during transmission.

**a.** Transmit classified information and material according to the security classification level, as specified in AR 380-5.

**(1)** Transmit classified material by—

**(a)** Approved courier services.

**(b)** U.S. military personnel.

**(c)** Government employees.

**(d)** DOD contractors.

**(e)** U.S Postal Service.

**(f)** Electronic transmission over secure lines using encryption.

**b.** Ensure transmission has been authorized in writing by the appropriate contracting officer.

**Performance Steps**

**c.** Ensure that TOP SECRET and SECRET classified information remain in constant custody and protection of the courier at all times.

**9.** Safeguard classified material when in storage.

**a.** Store classified information under conditions adequate to prevent unauthorized access.

**b.** Store typewriter ribbons, computer disks, notes, and similar materials which are classified or used to process classified information, in a locked General Services Administration (GSA) approved security container, when not under the personal control and observation of an authorized person.

**c.** Follow correct procedures when locking material in a GSA approved security container:

**(1)** Complete correctly SF 702 with the proper date, time, and initials.

**(2)** Turn the "OPEN/CLOSED" sign on the front of the security container to the "CLOSED" position.

**(3)** Ensure that another person checks the container to make sure that it is locked and initials the SF 702.

**10.** Use the following procedures when discovering an open or unattended security container.

**a.** Keep the container or area under guard or surveillance.

**b.** Notify one of the persons listed on Part 1 of SF 700 attached to the inside of the security container drawer. If one of these individuals cannot be contacted, notify the duty officer, security manager, or other appropriate official.

**11.** Report possible compromise of classified information.

**a.** List individual responsibilities in cases of suspected compromise:

**(1)** Any person having knowledge of the loss or possible compromise of classified information will immediately report it to the security manager, S2, or commanding officer.

**(2)** Any person who discovers classified information out of proper control will take custody of the information, safeguard it, and immediately report to the security manager, S2, or commanding officer.

**(3)** The security manager, S2, or commanding officer will initiate a preliminary inquiry to determine the circumstances surrounding the loss or possible compromise of classified information and establish one of the following:

**(a)** The loss or compromise of classified information did not occur.

**(b)** The loss or compromise did occur but the compromise reasonably could not be expected to cause damage to the national security.

| Performance Steps |
|---|

**(c)** The loss or compromise of classified information did occur and that the compromise reasonably could be expected to cause damage to the national security.

**(4)** The security manager or S2 will report the circumstances of the compromise to—

**(a)** HQDA (DAMI-CIS) when dealing with SECRET and TOP SECRET information.

**(b)** The commander when dealing with CONFIDENTIAL information.

**(5)** A further investigation may be initiated if warranted.

**12.** Use proper procedures when destroying classified information.

**a.** Destroy classified information to preclude recognition or reconstruction. Destroy classified information by burning, melting, chemical decomposition, pulping, pulverizing, crosscut shredding, or mutilation.

***Note***: Burning is the preferred method of destroying classified information.

**b.** Complete DA Form 3964 for TOP SECRET information. Records shall be dated and signed at the time of destruction.

---

**Evaluation Preparation:** *Setup*: Provide the Soldier with the materials listed in the conditions statement. Tell the Soldier to protect the classified material.

***Note to the Trainer:*** If this task is to be evaluated during training, prepare and provide the Soldier with material which is classified "FOR TRAINING PURPOSES ONLY." Ensure that the material is marked to this effect. The Soldier must get a GO on all items to receive a GO on this task.

| Performance Measures | GO | NO GO |
|---|---|---|
| **1.** Identified classification of classified material. | —— | —— |
| **2.** Located and read the declassification and downgrading instructions. | —— | —— |
| **3.** Marked documents with all required identification and markings, or corrected incorrect markings. | —— | —— |
| **4.** Restricted access to classified information: | —— | —— |
| **a.** Determined the requirement for access before releasing classified material to other persons: | | |
| **(1)** Determined the need-to-know. | | |
| **(2)** Determined security clearance. | | |
| **b.** Prevented unauthorized persons from gaining access to classified material. | | |
| **(1)** Kept all classified documents under constant observation when removed from storage. | | |

| Performance Measures | GO | NO GO |
|---|---|---|
| **(2)** Kept documents face down or covered when not in use. | | |
| **(3)** Took appropriate measures when uncleared personnel entered the work area. | | |
| **(4)** Placed appropriate cover sheets on classified material. | | |
| **5.** Safeguarded classified information and material. | ___ | ___ |
| **a.** Safeguarded classified information when temporarily departing the work area. | | |
| **b.** Prevented display of classified information in public places. | | |
| **c.** Used correct procedures when creating classified working papers. | | |
| **6.** Complied with security regulations when discussing classified information. | ___ | ___ |
| **a.** Refrained from discussing classified and/or sensitive information on an unsecured telephone. | | |
| **b.** Used correct security procedures when presenting a classified briefing. | | |
| **c.** Used proper controls when using automated information systems. | | |
| **7.** Protected classified information during transport. | ___ | ___ |
| **8.** Identified reference delineating correct transmission procedures. | ___ | ___ |
| **9.** Locked classified material in an appropriate security container (when not in use). | ___ | ___ |
| **a.** Locked and/or checked the security container. | | |
| **b.** Initialed and dated the SF 702. | | |
| **c.** Turned the sign to the "CLOSED" position. | | |
| **10.** Took appropriate action upon discovery of possible compromise of classified information. | ___ | ___ |
| **11.** Completed DA Form 3964, and determined the best destruction method for documents of various classifications. | ___ | ___ |
| **a.** Listed three methods of destruction. | | |
| **b.** Identified the two standards of destruction. | | |

**Evaluation Guidance:** Refer to chapter 1, paragraph 1-4 b (6).

**References**

**Required:** AR 380-5, DA Form 3964, SF 700, SF 702, SF 703, SF 704, SF 705, and USOP

**Related:**

**SUBJECT AREA 24: ENEMY PERSONNEL**

## 191-377-4203

### Supervise the Establishment and Operation of a Roadblock/Checkpoint

**Conditions:** You are given an operation order (OPORD), required personnel, a combat load, the unit standing operating procedure (SOP), maps and overlays, signal operating instructions (SOI), communications equipment, and materials to use as obstacles and guide signs.

**Standards:** Establish the roadblock, checkpoint, and holding areas according to the mission requirements. Establish and maintain communications according to the unit SOP. Ensure that the team or squad members correctly perform their assigned duties.

**Performance Steps**

**1.** Select a location for the roadblock within the assigned area.

**a.** Ensure that the location provides good cover and concealment for the team.

**b.** Position the roadblock so that unauthorized vehicles or enemy personnel cannot bypass it.

**c.** Place the roadblock at an intersection so that drivers can change to another route with little delay if the roadblock is used to close off a road.

**d.** Select a location where there is room for drivers to turn the vehicles around easily if there is no intersection.

**e.** Select a location for the roadblock where it cannot be seen until the drivers have passed all possible turnoffs if the roadblock is to be used with a checkpoint.

*Note*: Notify higher headquarters of your arrival time and your exact location.

**2.** Establish defensive positions.

**a.** Select fighting positions for the automatic weapons.

**b.** Ensure that the emplacement of the weapons provides security and overwatch for the roadblock and the holding area.

**Performance Steps**

**c.** Ensure that the team's vehicle is placed in a covered and concealed location near the team's or squad leader's position.

**d.** Direct the team or squad to camouflage the weapon and vehicle if natural cover and concealment are not available in step 2b and step 2c.

**3.** Establish communications according to the unit SOP.

**4.** Supervise the establishment of the roadblock.

**a.** Direct the barricading of the road, road shoulders, and ditches to channel the passing traffic.

**b.** Direct the placement of signs to warn drivers that a roadblock is ahead.

**c.** Ensure that there is adequate lighting for drivers to see the roadblock if the roadblock is set up during limited visibility.

**5.** Select an appropriate checkpoint and holding area within the assigned area based on the purpose of the checkpoint.

**a.** Place the checkpoint at the entrance of the controlled route if the purpose is to check convoys for authorization to use the route.

**b.** Place the checkpoint just over a hill or around a curve if the purpose is to check the cargo or to spot check vehicle traffic.

**c.** Select a location for the holding area.

**(1)** Ensure that the vehicles can disperse if the tactical situation demands it.

**(2)** Ensure that there is easy access to and from the roadway.

**(3)** Ensure that the surface of the area is firm enough to hold the weight of the vehicles.

**(4)** Ensure that the area is large enough to allow the vehicles to be covered and concealed from air and ground observation.

**(5)** Ensure that the area is easy to defend.

**6.** Supervise the establishment of a checkpoint.

**a.** Perform the procedures in step 2 to establish defensive positions.

**b.** Direct the team or squad in establishing the checkpoint where it will be hidden from distant view, unless used at the entrance to a controlled route.

**c.** Direct the outlining of immediate approach lanes with engineer tape, instructional signs, debris, trees, or anything that can be used as an obstacle.

**d.** Direct the placement of signs and flares to warn drivers that a checkpoint is ahead.

**e.** Use a roadblock to stop or channel traffic, if necessary.

| Performance Steps |
| --- |

**7.** Supervise the establishment of the holding area.

**a.** Perform the procedures in step 2 to establish defensive positions.

**b.** Instruct the military police to direct vehicles into the holding area so that the first vehicle in is able to be the first vehicle out.

**c.** Ensure that vehicles are parked facing the exit.

**d.** Notify the headquarters when the checkpoint is operational.

**e.** Ensure that the personnel manning the holding area are briefed on special instructions.

**8.** Supervise the operation of a roadblock and a checkpoint. Supervise a—

**a.** Roadblock.

**(1)** Assign each team or squad member to a position.

**(2)** Explain each member's duties and responsibilities.

**(3)** Maintain communications with the team or squad members and headquarters.

**(4)** Oversee the operation of the roadblock to ensure that the vehicles are stopped, checked for appropriate documents and authorizations, and searched according to the specifications of the mission.

**b.** Checkpoint.

**(1)** Perform steps 8a (1), (2), and (3).

**(2)** Oversee the operation of the checkpoint.

**(a)** Ensure that the convoys are checked for movement credits according to the highway traffic division (HTD) and that drivers who are lost are directed to their destinations if the checkpoint is at the entrance to a controlled route.

**(b)** Ensure that vehicles are checked for movement credits according to the HTD. Check the manifest papers against the actual load when checking cargo vehicles if the checkpoint is on a main supply route.

**(c)** Remind the team or squad members to be suspicious of military equipment, supplies, or weapons transported in civilian vehicles.

---

**Evaluation Preparation:** *Setup*: Schedule a field training exercise with vehicle traffic so that the Soldier can establish and supervise a roadblock, checkpoint, and holding area. Identify a general location. Evaluate the Soldier's ability to select a specific location that offers cover and concealment.

*Brief Soldier:* Tell the Soldier that he/she will be evaluated as a team or squad leader. Tell the Soldier that contact with the enemy is to be expected and that all vehicles must be stopped and checked.

| Performance Measures | GO | NO GO |
|---|---|---|
| **1.** Selected the location for the roadblock within the assigned area. | —— | —— |
| **2.** Established defensive positions. | —— | —— |
| **3.** Established communications according to the unit SOP. | —— | —— |
| **4.** Supervised the establishment of the roadblock. | —— | —— |
| **5.** Selected an appropriate checkpoint and holding area within the assigned area based on the purpose of the checkpoint. | —— | —— |
| **6.** Supervised the establishment of a checkpoint. | —— | —— |
| **7.** Supervised the establishment of the holding area. | —— | —— |
| **8.** Supervised the operation of a roadblock and a checkpoint. | —— | —— |

**Evaluation Guidance:** Refer to chapter 1, paragraph 1-4 b (6).

**References**

**Required:** FM 3-19.4

**Related:**

# 191-377-4250

## Supervise the Processing of Detainees at the Point of Capture

**Conditions:** You are given Soldiers with individual equipment (assigned weapons, disposable restraints [wrist and ankle], soft cloths, and permanent markers), an interpreter (when available), the rules of engagement (ROE)/rules for the use of force (RUF), AR 190-8, DD Forms 2745 (*Enemy Prisoner of War [EPW] Capture Tag*), DA Form 4137 (*Evidence/ Property Custody Document*), STP 19-31B1-SM, STP 19-31B24-SM-TG, first aid supplies, communications equipment, and access to the Geneva Conventions and the Hague Convention.

**Standards:** Ensure that Soldiers follow the proper procedures when searching detainees. Ensure that Soldiers tag detainees using a DD Form 2745, document the search using a DA Form 4137, and link both by the DD Form 2745 number. Ensure that Soldiers keep detainees silent and segregated. Safeguard detainees according to U.S. policy and Army values. Enforce ROE/RUF during escape attempts. Correct and report inappropriate treatment and detainee noncompliance.

| Performance Steps |
|---|
| ***Note:*** The best way to ensure readiness for detainee operations is proper planning. Prior to executing missions, gather the latest intelligence capture estimates, if possible. Brief Soldiers, review the ROE/RUF, conduct rehearsals, and ensure that all the required equipment and forms for detainee processing are available before executing the mission. |

## Performance Steps

**1.** Disarm and secure the detainees.

***Note:*** Depending on the circumstances of capture (such as the battle outcome, surrender, and the presence of refugees), tempers may be high and communications may be difficult. Verbal commands or hand signals may have to be reinforced and given several times, especially when detainees do not understand English.

**a.** Upon capture or surrender of combatant forces, direct them to—

**(1)** Place their weapons on the ground, raise their hands above their heads, and step back 5 paces.

**(2)** Raise their shirts, shake their clothing loose, and turn 180 degrees in place to allow a visual check for other weapons or explosives.

***Note:*** In the instance of a suicide bomber, the first responsibility is the safety of fellow Soldiers and bystanders. Attempt to remove all bystanders, detainees, and Soldiers from the possible blast zone of a suicide bomber. If that is not feasible, refer to the local ROE for direction.

**b.** Position guards to ensure that they have an unobstructed line of fire.

**c.** Direct the designated team to move forward to check wounded or dead detainees.

**(1)** Ensure that the team members disarm, restrain, and remove wounded detainees to a separate location and position them under guard.

**(2)** Ensure that the team members search wounded detainees. Approach wounded detainees with caution, with the weapon drawn on the detainee, until the detainee's weapon is out of his/her reach.

**(3)** Direct a team member and/or a combat lifesaver to provide medical assistance for wounded detainees under the supervision of guards.

**(4)** Segregate wounded detainees from nonwounded detainees.

**d.** Secure weapons and remove weapons on the ground to a safe location in view of the guards. Document the owner of each weapon. Soldiers should not cross between a detainee and the armed guard in case the detainee tries to attack the Soldier.

***Note:*** Detainee weapon documentation should be done when the situation allows. Under stressful situations, the DD Form 2745 number can be annotated on the stock of the weapon with a permanent marker and the DA Form 4137 completed later when in a safe area.

**2.** Organize the Soldiers and prepare to field-process detainees.

**a.** Establish an outer perimeter to provide force protection for Soldiers and detainees.

**b.** Notify higher headquarters of the capture.

**c.** Establish a search team, which will consist of a Soldier to conduct the search, a Soldier to provide security for the Soldier conducting the search, and an interpreter, if available. Each search team should only process one detainee at a time.

**Performance Steps**

**d.** Establish two internal guard teams to secure the detainees that have not been searched and the detainees that have been searched. They will escort detainees to and from the search team and segregation areas.

**e.** Ensure that all search teams have detainee field-processing equipment (for example, disposable restraints, cloth, permanent markers, DD Form 2745s, and DA Form 4137s).

**f.** Assume a position to maintain situational understanding of the entire operation. Attention cannot be focused only on the actions of one searcher because of responsibilities for the safety, security, and actions of Soldiers providing force protection security, guarding detainees awaiting the search process, and guarding detainees that are through the search process.

**3.** Direct search teams to process detainees.

**a.** Position members of each search team.

**(1)** Position the guard to ensure that he/she has a clear line of sight to the detainee.

**(2)** Position the interpreter to best support the searcher (normally to the guard's rear flank).

**(3)** Position the searcher to conduct the search and instruct the guard and interpreter as necessary.

**b.** Ensure that Soldiers tag detainees according to the proper procedures.

**(1)** Ensure that Soldiers initially confiscate all items from detainees (some items will be returned immediately following the search).

***Note:*** Currency will only be confiscated on the express order of a commissioned officer (for example, through an operations order, a fragmentary order, or a verbal order) per AR 190-8. Soldiers must notify a commissioned officer when currency is found. If a decision is made to confiscate it, it must be accounted for on a DA Form 4137.

**(2)** Have the Soldiers that performed the search fill out a DD Form 2745 when the search has been completed.

**(3)** Instruct Soldiers to tag the detainee and the confiscated items, using all three copies of the DD Form 2745.

**(a)** Attach part A to the detainee.

**(b)** Retain part B for official records.

**(c)** Attach part C to confiscated items.

***Note:*** The best time to complete the DD Form 2745 is immediately following the search. However, if speed is required, you may complete it later. If you have enough Soldiers, you may want to establish another team to tag the detainees.

**(4)** Direct Soldiers to complete a DA Form 4137 and secure and document confiscated items if the situation permits.

**Performance Steps**

*Note:* Ensure that Soldiers mark confiscated items that will be retained or destroyed (such as weapons, personal items, and investigatory/evidentiary items) for identification if time permits. Use a plastic bag or similar container to identify, collect, and temporarily store items that cannot be individually marked because of insufficient time due to imminent attack or any other emergency. Mark property according to STP 19-31B24-SM-TG.

**4.** Segregate detainees at the point of capture based on mission necessity.

**a.** Base initial segregation on information surrounding the capture events.

**b.** Segregate detainees to the greatest extent possible as more detailed information is known and resources and conditions allow.

**5.** Ensure that Soldiers maintain silence among detainees.

**a.** Ensure that Soldiers who are guarding the detainees maintain detainee silence. All guards should be alert for detainee leaders trying to give orders and for any attempts to plan an escape.

**b.** Segregate uncooperative detainees to minimize their effect on others if they remain uncooperative.

**6.** Ensure that Soldiers safeguard detainees from the moment of capture until they are released or repatriated.

**a.** Correct and report any inappropriate treatment immediately.

**b.** Ensure that Soldiers respond to escape attempts according to the ROE and RUF.

**7.** Update higher headquarters on the situation after all the detainees have been processed. Include—

**a.** The date, time, and location of capture.

**b.** The total number of detainees, their categories (if known), and all DD Form 2745 numbers.

**c.** Any confiscated items of intelligence value.

**d.** Any acts or allegations of inhumane treatment or abuse.

**e.** Any detainees requiring medical assistance.

**8.** Coordinate with higher headquarters the evacuation of detainees.

*Note:* Supervisors must ensure that detainees are transported quickly to the detainee collection point or detainee holding area based on security, operational conditions, and available transportation.

**a.** Obtain the location of the designated collection point, and execute the mission according to STP 19-31B24-SM-TG.

| Performance Steps |
| --- |
| **b.** Coordinate the date, time, and location of custody transfer if another unit will transport the detainees. |
| **c.** Request additional resources to include food, water, and support along the escort route if they are not already factored by headquarters. |

**Evaluation Preparation:** *Setup:* Provide the Soldier with four or more personnel to act as detainees and guards, materials to be used as detainee props, and the materials and references listed in the conditions statement.

*Brief Soldier:* Tell the Soldier that he/she will be evaluated on his/her ability to supervise the processing of detainees and equipment.

| Performance Measures | GO | NO GO |
| --- | --- | --- |
| **1.** Disarmed and secured the detainees. | —— | —— |
| **2.** Organized the Soldiers and prepared to field-process detainees. | —— | —— |
| **3.** Directed search teams to process detainees. | —— | —— |
| **4.** Segregated detainees at the point of capture based on mission necessity. | —— | —— |
| **5.** Ensured that Soldiers maintained silence among detainees. | —— | —— |
| **6.** Ensured that Soldiers safeguarded detainees from the moment of capture until they were released or repatriated. | —— | —— |
| **7.** Updated higher headquarters on the situation after all the detainees were processed. | —— | —— |
| **8.** Coordinated the evacuation of detainees with higher headquarters. | —— | —— |

**Evaluation Guidance:** Refer to chapter 1, paragraph 1-4 b (6).

**References**

**Required:** AR 190-8, DA Form 4137, and DD Form 2745

**Related:** FM 3-19.40

## 191-377-4252

### Supervise the Escort of Detainees

**Conditions:** In a field environment, you are given a mission to escort detainees from one location to another, fully equipped Soldiers, completed DD Forms 2745 (*Enemy Prisoner of War [EPW] Capture Tags*), DA Form 4137 (*Evidence/Property Custody Document*), blank DD Form 2708s (*Receipt for Inmate or Detained Person*), the rules of engagement (ROE)/rules for the use of force (RUF), first aid supplies, communications equipment, an interpreter (if available), confiscated property, and a known mode of transportation.

**Standards:** Brief the Soldiers on the mission. Supervise the safe escort of all assigned detainees from one location to another. Maintain accountability for the detainees and confiscated property.

| **Performane Steps** |
|---|

**1.** Plan the detainee escort mission.

**a.** Confirm the mission route, the date, the time, the number of detainees to be transferred, the number of guards, and the mode of transportation.

**b.** Plan the convoy escort based on a risk assessment of detainee status, security, force protection, and resource requirements.

**c.** Task organize your personnel to complete the mission. Supervisors must—

**(1)** Consider the guard-to-detainee ratio based on the threat, the nature of the detainees, the route, the timing, and the external risks.

**(2)** Identify Soldiers for the guard force, train and/or brief them, and rehearse roles and responsibilities.

**(3)** Identify Soldiers for the security force and train and/or brief them on their duties and the ROE/RUF.

**d.** Conduct a route and/or map reconnaissance.

**(1)** Consider the likelihood of the presence of sympathizers, hostile local nationals, or terrorists and their use of improvised explosive devices (IEDs).

**(2)** Determine the location of military police units or other security units along the route.

**(3)** Determine the rest stops.

**(4)** Determine any additional resources required for the convoy, including food, water, and sustainment items.

**2.** Brief Soldiers and rehearse the mission, when possible.

**a.** Ensure that movement procedures are briefed to all escort Soldiers with an interpreter (if available), including any support personnel (such as drivers and medics) to ensure that they understand what to do during—

**(1)** A convoy.

**(2)** Planned stops.

**(3)** Unexpected stops.

**(4)** An IED incident.

**(5)** An ambush by an aircraft and/or a sniper.

**(6)** Detainee-initiated disturbances.

**(7)** Detainee and/or local population interaction.

**b.** Rehearse the mission when appropriate.

**3.** Prepare the detainees for transport.

**a.** Direct the Soldiers to search the detainees for any weapons or contraband that may have been missed by the initial search.

**Performane Steps**

**b.** Prepare a manifest or roster listing all of the detainees that will be escorted. Account for detainees using a locally produced manifest. Use any appropriate method available, such as a computer program or a pencil and paper, to create a manifest that includes the following minimum required information:

**(1)** Name (if known).

**(2)** DD Form 2745 number.

***Note:*** The DD Form 2745 number is the only number used to account for a detainee until an internment serial number (ISN) is assigned at a theater internment facility (TIF).

**(3)** The capturing unit.

**(4)** The vehicle in which the detainees will be transported.

**(5)** The destination.

**c.** Inventory detainees by using the DD Form 2745 numbers. Ensure that each detainee has been tagged using a DD Form 2745 and that Part A is hanging visibly from the detainee. If this is not complete, ensure that the releasing unit prepares the DD Form 2745 before accepting custody of the detainees. If your unit was the capturing unit, you or your Soldiers must prepare the DD Forms 2745.

**(1)** Inventory confiscated item bundles using DA Form 4137s.

**(a)** Ensure that all confiscated equipment and material is accounted for, linked to the DD Form 2745 number, and tagged with DD Form 2745, Part C.

**(b)** Ensure that the releasing unit prepares the proper documentation before accepting the property if the confiscated equipment and material has not been tagged.

**(2)** Prepare the proper accountability documents, or have your Soldiers prepare them if your unit was the capturing unit.

**d.** Brief detainees on the planned movement. Use an interpreter (when available) to give clear, brief instructions in their own language, if possible. Give no more information than is necessary to the detainees. As a minimum, brief them on the following:

**(1)** The actions to take upon hearing the command "Halt."

**(2)** The requirement to remain silent at all times.

***Note:*** It may become necessary to muffle and/or blindfold unruly detainees. If required and authorized, silence uncooperative detainees by muffling them using a soft, clean cloth tied around their mouth and fastened at the back of the head. Do not harm the detainees or affect the detainees' ability to breathe. Do not use duct tape or other adhesives, place cloth or objects inside the mouth, or use physical force to silence detainees. Instruct guards to periodically check detainees to ensure that they are not harmed. Muffle detainees only for as long as needed.

**(3)** The actions they are to take during an emergency, such as a delay or a crash.

**Performane Steps**

(4) The hand-and-arm signals used to direct detainee movement.

(5) The implications of, and the guard force response to, escape attempts.

**e.** Ensure that detainees are properly restrained during transport.

(1) Do not daisy chain detainees together or fasten them to a fixed object during movement unless it is a military necessity or commanded by a higher authority. Vehicle collisions could severely injure restrained limbs.

(2) Ensure that detainees are always wearing a seat belt while onboard an aircraft.

(3) Use blindfolds as appropriate to provide operations security.

**f.** Maintain segregation to the maximum extent possible.

**4.** Ready the detainees for movement or transport.

**a.** Ready the detainees for movement by foot.

(1) Put the detainees in close-column formation.

***Note:*** Disposable restraints can remain fastened behind the detainees' backs, but the supervisor has discretion to place disposable restraints to the front when escorting detainees by foot. Factors to consider for this decision should include the type of terrain, distance, and length of travel.

***Note:*** Do not blindfold detainees for movement by foot.

(2) Place escort vehicles at the front and rear of the columns.

(3) Transport supplies in vehicles.

(4) Designate a vehicle in the rear of the formation for chase and/or straggler control.

(5) Designate a separate vehicle in the rear for medical support.

(6) Designate walking guards on the flanks.

***Note:*** The number of guards required will be based on mission, enemy, terrain and weather, troops and support available, time available, and civil considerations (METT-TC) and the supervisor's discretion.

***Note:*** Ensure that Soldiers search the transport vehicles for contraband both before and after each escort.

**b.** Load the detainees for transport by wheeled vehicle.

(1) Search each vehicle to ensure that no contraband or weapons are present.

(2) Position guards at the front and rear of each vehicle.

(3) Place the escort security vehicles at the front and rear of the columns.

(4) Designate the vehicles in the rear to provide flank security and chase.

(5) Ensure that Soldiers assist restrained detainees on and off vehicles.

**c.** Load the detainees for transport by rail.

(1) Search each car to ensure that there are no weapons or contraband present.

| Performane Steps |
| --- |

***Note:*** Blindfolds may not be necessary, based on the mode of transportation.

***Note:*** Position the guards according to the METT-TC, the number of available guards, and the type of railcar used. Passenger cars are preferred over freight cars.

**(2)** Assist restrained detainees on and off cars.

**d.** Load the detainees for transport by air.

**(1)** Search the aircraft before and after the mission to ensure that there are no contraband or weapons present.

***Note:*** There may be a need to place disposable restraints on detainees' ankles. This is based on guidance from the aircraft commander and/or crew chief.

**(2)** Ensure that Soldiers assist restrained detainees on and off the aircraft.

**(3)** Ensure that Soldiers follow the instructions of the aircraft commander and/or crew chief for flight operations rules (such as requirements for seat belts, hearing protection, or special restraints).

**5.** Supervise and monitor the execution of the mission.

**a.** Ensure that the guards remain vigilant at all times. Monitor guards to ensure that they—

**(1)** Maintain weapon control and muzzle awareness.

**(2)** Maintain detainee silence and segregation.

**(3)** Restrain the detainees appropriately and check restraints periodically.

**(4)** Use a guard for over watch when in contact with detainees.

**(5)** Eliminate and/or reduce any contact between the detainees and the local population.

***Note:*** Detainees will identify and exploit signs of guard boredom, complacency, and fatigue.

**b.** Prepare for actions at the planned stops and anticipate actions at unexpected stops.

**c.** Maintain continuous contact with higher headquarters and execute the mission according to the local standing operating procedure (SOP).

**d.** Direct response actions during emergencies.

**e.** Correct, report, and document any allegations of detainee abuse.

**6.** Coordinate with the gaining unit assuming custody of the detainees upon arrival at the designated location (detainee collection point [DCP], detainee holding area [DHA], or TIF).

**a.** Continue to provide security until properly relieved of custody.

**b.** Review the detainee escort manifest with assigned facility personnel.

| Performane Steps |
|---|

**c.** Conduct a joint inventory of all detainees, pertinent documents, and confiscated items using the escort manifest, to include—

**(1)** DD Form 2745 (Parts A, B, and C).

**(2)** DA Form 4137.

**(3)** Medical files or information if available.

**d.** Provide any additional information regarding the circumstances of capture.

**e.** Provide a unit point of contact for any necessary follow-up.

**f.** Have the escort supervisor sign for the release of confiscated items on the DA Form 4137 when the joint inventory has been completed. The gaining authority signs for receipt of confiscated items on the DA Form 4137, prepares and signs the DD Form 2708, and returns the completed form to the escort supervisor.

**g.** Transfer custody of the detainee, respective documents, and confiscated items.

**h.** Brief facility personnel on all issues or information pertaining to the detainees or their conduct during the escort.

**7.** Notify higher headquarters of the total number of detainees transferred using DD Form 2745 numbers.

---

**Evaluation Preparation:** *Setup:* Provide the Soldier with four or more personnel to act as detainees, materials to be used as detainee props, Soldiers as guards, required documentation, and a known mode of transportation.

*Brief Soldier:* Tell the Soldier that he/she will be evaluated on his/her ability to supervise the escort of detainees.

| Performance Measures | GO | NO GO |
|---|---|---|
| **1.** Planned the detainee escort mission. | ___ | ___ |
| **2.** Briefed the Soldiers and rehearsed the mission, when possible. | ___ | ___ |
| **3.** Prepared the detainees for transport. | ___ | ___ |
| **4.** Readied the detainees for movement or transport. | ___ | ___ |
| **5.** Supervised and monitored the execution of the mission. | ___ | ___ |
| **6.** Coordinated with the gaining unit assuming custody of the detainees upon arrival at the designated location (DCP, DHA, or TIF). | ___ | ___ |
| **7.** Notified higher headquarters of the total number of detainees transferred using DD Form 2745 numbers. | ___ | ___ |

**Evaluation Guidance:** Refer to chapter 1, paragraph 1-4 b (6).

**References**

**Required:** DA Form 4137, DD Form 2708, and DD Form 2745

**Related:** AR 190-8 and FM 3-19.40

## 191-378-4303

### Supervise a Riot/Crowd Control Operation with a Squad- Sized Element

**Conditions:** You are given a riot or crowd control situation and a squad of Soldiers fully equipped with riot control gear, FM 3-19.15, and STP 19-31B1-SM.

**Standards:** Based on the situation, select the best formation to control or disperse the crowd. Correctly form the squad into the formation selected. Move the squad to disperse or move the crowd and reassemble the squad after the crowd has been controlled or dispersed.

**Performance Steps**

**1.** Analyze the mission requirements.

*Note*: Squad-sized formations are limited based on the number of squad members. A leader should employ a unit-sized formation to fit the circumstances. Squad formations are the smallest formations used and are usually employed to cover a main formation's flanks, such as side streets.

**a.** Determine the size of the area to maneuver the unit through.

**b.** Determine the approximate number of people in the crowd.

**c.** Determine if the mission goal is to—

**(1)** Move the crowd.

**(2)** Make apprehensions.

**(3)** Perform an extraction.

**d.** Determine the type of formation to be used, based on the mission.

**(1)** Direct the squad in forming as a line, if required (figure 191-378-4303-1).

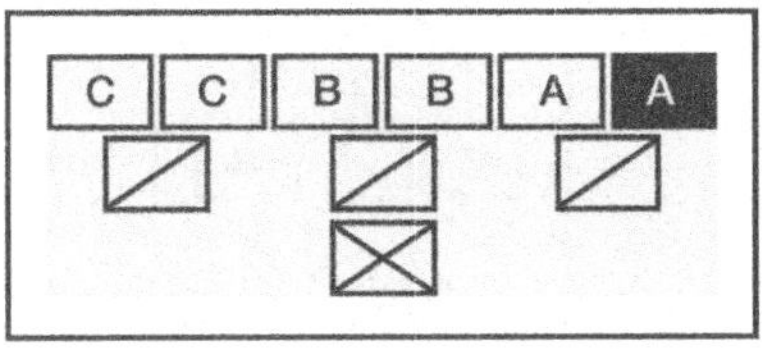


**Figure 191-378-4303-1. Squad line formation**

*Note*: This formation is used to push a crowd in a straight line backwards.

**Performance Steps**

**(2)** Direct the squad in forming a squad echelon, right or left, if required (figure 191-378-4303-2).

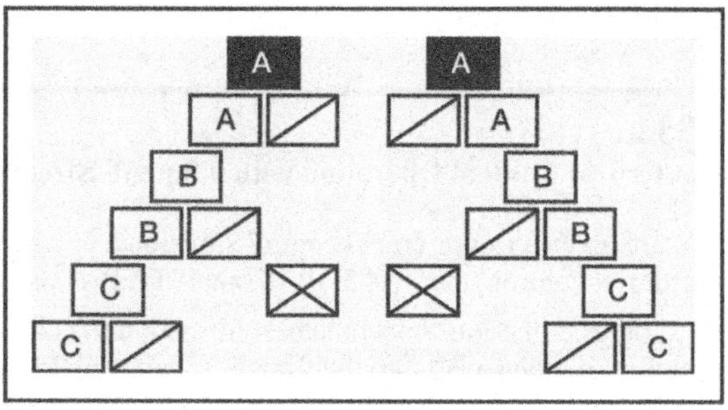


**Figure 191-378-4303-2. Squad echelon formations**

*Note*: This formation is used offensively to turn crowds in either open or built-up areas and to move them away from buildings, fences, or walls.

**(3)** Direct the squad in forming a squad wedge, if required (figure 191-378-4303-3).

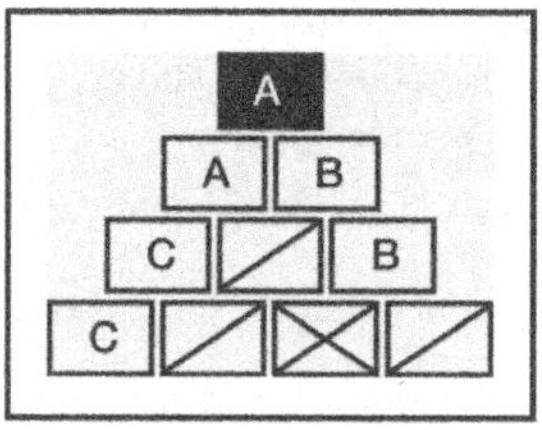


**Figure 191-378-4303-3. Squad wedge formation**

*Note*: This formation is used as an offensive formation to penetrate or split crowds.

**(4)** Direct the squad in forming a diamond or circle, if required (figure 191-378-4303-4).

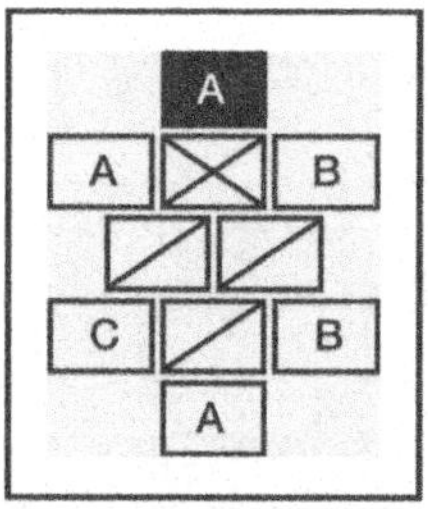


**Figure 191-378-4303-4. Squad diamond or circle formation**

*Note*: This formation is used during extraction or recovery operations to penetrate crowds and extract someone or something from that crowd. The decision to use this formation is based on mission, enemy, terrain and weather, troops and support available, time available, civil considerations (METT-TC) and the ability of the squad to perform.

**e.** Identify any mission support requirements, to include—

**(1)** A back-up support squad.

**(2)** Vehicle support.

**Performance Steps**

(3) Video photographer support.

(4) Communications support.

(5) Medical support.

f. Complete a risk assessment.

g. Identify potential threats to the rear of the formation once crowd control movements have been completed.

h. Identify the team or operation exit strategy.

2. Identify any nonlethal weapons (NLWs) requirements, to include—

a. A NLWs Capability SE Kit.

b. A 12-gauge shotgun.

c. An M203 grenade launcher.

3. Identify any lethal overwatch team requirements.

*Note*: Lethal overwatch teams are assigned based on the METT-TC, mission risk factors, availability, and the environment in which the mission is to be performed. For example, a hostile fire zone would require the most support from lethal overwatch teams. Examples of when lethal overwatch teams are authorized to employ lethal force are situations involving—

- Self-defense and the defense of others based on imminent danger of death.
- Assets of national security that are in danger.
- Assets inherently dangerous, such as explosives or chemical or nuclear materials.

a. Lethal overwatch teams must be briefed on their ROE and must have communication with the movement leader.

b. Sniper cover support to the formation movement, as appropriate.

c. Vehicle crew serve support to the formation movement, as appropriate.

d. Countersniper support to the formation movement, as appropriate.

4. Organize the squad and support for formation.

a. Brief the squad and all support teams on—

(1) The mission goals.

(2) Intelligence relevant to the mission.

(3) Contingencies.

(4) The NLWs to be used and their location in the formation.

(5) The NLWs assignments of squad members.

**Performance Steps**

**b.** Form the squad in a single column behind you with the base member directly behind you and the assistant squad leader at the end.

**c.** Direct squad members to count off from the front to the rear, with you as number 1 and the base member as number 2.

**d.** Tell the squad members which weapon positions to use.

***Note***: Each position has a specific use.

**(1)** Use the safe-port position when making a show of force.

**(2)** Use the safe-guard position for semireadiness.

**(3)** Use the on-guard position for complete readiness and when troops are in contact with a group showing any kind of resistance or hesitance to withdraw.

**5.** Move the formation from the staging area to the mission area.

**a.** Face the squad and give the command, "Squad line formation."

**b.** Give hand-and-arm signals (simultaneously with the command) by raising the arms straight out, horizontal to the shoulders, hands extended, and palms down.

**c.** Point to where you want the squad to assemble and complete the command, "Move," and ensure that the base member moves where indicated and that the other squad members form according to their number.

**d.** Take your position behind the formation.

***Note***: The squad leader may use one or all of steps 6, 7, and 8 based on the situation and the formation selected and/or required.

**6.** Direct the squad in forming a squad echelon right (or left), if required.

**a.** Face the squad and give the command, "Squad echelon right (or left)," while simultaneously giving the hand-and-arm signals by extending one arm 45 degrees above the horizon and the other arm 45 degrees below the horizon with arms and hands extended.

***Note***: When facing the squad, the upper arm shows the direction of the echelon.

**b.** Ensure that the base member moves where indicated and that the rest of the squad aligns themselves with the base member (one pace behind the member in front and one pace to the right or left, depending on the echelon).

**c.** Point to where the squad should assemble and complete the command, "Move."

**d.** Assume the position behind the formation.

**Performance Steps**

**7.** Direct the squad in forming a squad wedge, if required.

**a.** Face the squad and give the command, "Squad wedge," while simultaneously giving the hand-and-arm signals by extending both arms down and to the sides at an angle of 45 degrees below the horizon with arms and hands extended and palms down.

**b.** Ensure that the base member moves where indicated and that the odd-numbered members align themselves behind the base member (one pace to the left and one pace to the rear of the preceding member) and that the even-numbered members align themselves behind the base member (one pace to the right and one pace to the rear of the preceding member).

**c.** Point to where the squad will assemble and complete the command, "Move."

**8.** Direct the squad in forming a diamond (or circle), if required.

**a.** Face the squad and give the command, "Squad diamond (or circle), move," while simultaneously giving hand-and-arm signals by extending the right arm above the head and rotating it in a circle with the left arm at a 45 degree angle, pointing at the circling hand.

**b.** Ensure that the following are completed on the command of execution:

**(1)** The A team leader moves the baseman to the location indicated by the squad leader.

**(2)** The A team lines up to the left of the baseman, covering the 8 to 12 o'clock positions.

**(3)** The B team lines up to the right of the baseman, covering the 12 to 4 o'clock positions.

**(4)** The C team lines up covering the 4 to 8 o'clock positions.

**9.** Move the formation using the command, "Forward, march."

***Note***: This will begin the forward movement of the formation. Left and right formation movement adjustments will be made by using the command, "Echelon left" or "Echelon right." All other movement commands, such as "Mark time" or "Halt," can be used as appropriate.

**a.** Identify the flow of the crowd as the team advances and make commands to the team for formation movements to compensate for the crowd's movement and ensure that the squad achieves the goals assigned.

***Note***: The squad leader and squad members should be mindful of safety factors in formation movement—such as airborne projectiles, antagonistic individuals in the crowd who may need removal, possible incoming direct fire, and the need for peripheral view checks of squad members to the left and right.

**b.** Give the command, "Halt," once the movement of the crowd is complete according to the goals of the mission.

| Performance Steps |
|---|

**c.** Review the crowd dispersal and reevaluate the situation to ensure that the mission is complete.

**d.** Check peripherals, side, and rear to ensure that the squad is safe and secure.

**e.** Communicate intentions to remove the squad with any support teams before reassembling.

**10.** Reassemble the squad.

**a.** Ensure that the squad members hear the command and come to the safe-port position and that the base member faces the squad leader and the other squad members face toward the base member.

**b.** Take a position a sufficient distance to the rear of the squad and command, "Squad, assemble," while simultaneously raising the right hand in the air and making a circular motion.

**c.** Point to where the squad should assemble and give the command of execution "Move."

**d.** Ensure that the base member double-times to the designated spot, the other members of the squad follow the base member and that the squad forms in a column behind the base member and waits for further instruction.

**e.** Perform an after-action review with the squad and all mission support participants.

**(1)** Account for and check all participants to see if anyone requires medical attention.

**(2)** Perform weapons clearing functions, if not in a hostile theater of operations.

**(3)** Perform ammunition accountability procedures, if not in a hostile theater of operations.

**(4)** Review what went right.

**(5)** Review what could be improved from the completed operation.

**(6)** Release or move Soldiers according to the mission requirements.

**Evaluation Preparation:** *Setup*: Provide the Soldier with the materials and equipment listed in the conditions statement.

*Brief Soldier:* Tell the Soldier that he/she will be evaluated based on the performance measures.

| Performance Measures | GO | NO GO |
|---|---|---|
| **1.** Analyzed mission requirements. | —— | —— |
| **2.** Identified the NLWs requirements. | —— | —— |
| **3.** Identified the lethal overwatch team requirements. | —— | —— |
| **4.** Organized the squad and support for formation. | —— | —— |

| Performance Measures | GO | NO GO |
|---|---|---|
| **5.** Moved the formation from the staging area to the mission area. | ___ | ___ |
| **6.** Directed the squad in forming a squad echelon right (or left) formation, if required. | ___ | ___ |
| **7.** Directed the squad in forming a squad wedge formation, if required. | ___ | ___ |
| **8.** Directed the squad in forming a diamond (or circle) formation, if required. | ___ | ___ |
| **9.** Moved the formation using the command, "Forward, march." | ___ | ___ |
| **10.** Reassembled the squad. | ___ | ___ |

**Evaluation Guidance:** Refer to chapter 1, paragraph 1-4 b (6).

**References**

**Required:** FM 3-19.15 and STP 19-31B24-SM-TG

**Related:**

## 191-378-5315

### Supervise an Installation Access Control Point

**Conditions:** As a noncommissioned officer in charge (NCOIC), you are given an installation access control point, personnel, appropriate equipment, the current force protection condition (FPCON) threat level, the local standing operating procedure (SOP), AR 190-13, and AR 525-13.

**Standards:** Supervise an installation access control point according to the local SOP, AR 190-13, and AR 525-13.

**Performance Steps**

**1.** Ensure that the special guard orders include instructions for the following:

- **a.** Performing sign-in procedures.
- **b.** Maintaining access rosters (if applicable).
- **c.** Contacting off-site emergency vehicles.
- **d.** Maintaining on-site emergency vehicles.
- **e.** Processing of authorized identification card holders.
- **f.** Using the contact roster for key personnel.
- **g.** Using the map of the installation.

**Performance Steps**

**h.** Maintaining phone numbers for the key organizations of the installation.

**i.** Implementing random antiterrorism measures for FPCONs.

**j.** Using the proper levels of force.

**k.** Maintaining the list of personnel barred from the installation.

**2.** Conduct guard mount.

**3.** Ensure that initial communication checks, both telephonic (Class C line and direct line) and radio, are made with the military police desk.

**4.** Ensure that an equipment inventory is conducted and that discrepancies are reported before taking responsibility for the inventory.

**5.** Ensure that traffic control devices are employed according to each gate's special orders. Check the following devices:

**a.** Operational gate (electronic).

**b.** Barriers.

**c.** Directional signs.

**d.** Regulatory signs.

**e.** Warning signs.

**f.** Portable lighting.

**g.** Traffic cones.

***Note***: Equipment knocked over or damaged needs to be set up or replaced as soon as possible. These items are part of the working structure and help ensure that the mission operates smoothly and that officer safety and public safety are maintained.

**6.** Inspect the condition of the facilities (gate house and visitor's center) and surrounding areas.

**a.** Ensure that the facilities are functioning properly (doors, lighting, and rest rooms).

**b.** Ensure that the surrounding areas are properly maintained and free of debris and public hazards.

**c.** Ensure that the visitor's center and other public-access facilities present a neat and organized appearance.

**7.** Ensure that all gate discrepancies and facility deficiencies are reported or annotated for the record.

**a.** Annotate communication failures.

| Performance Steps |
|---|
| **b.** Annotate all work orders. |
| **c.** Report immediately all deficiencies that affect access control operations. |
| **8.** Review preliminary investigation reports generated on violations identified by access control personnel. |
| **9.** Ensure appropriate interaction by personnel with the public. |
| **a.** Prevent idle conversation impeding the flow of vehicle or foot traffic. |
| **b.** Ensure that personnel maintain military bearing (polite and courteous). |
| **10.** Ensure that immediate gate closure procedures are performed, according to the appropriate FPCON level (Normal, Alpha, Bravo, Charlie, Delta) and the local SOP. |
| **11.** Ensure that a gate equipment inventory is conducted and that the facility is inspected before the end of the shift. |

**Evaluation Preparation:** *Setup*: Provide the Soldier with copies of all references listed in the conditions statement.

*Brief Soldier:* Tell the Soldier to analyze the situation and correctly perform the necessary steps to supervise an access control point.

| Performance Measures | GO | NO GO |
|---|---|---|
| **1.** Ensured that the special guard orders included all required instructions. | ___ | ___ |
| **2.** Conducted the guard mount. | ___ | ___ |
| **3.** Ensured that initial communication checks, both telephonic and radio, were made with the military police desk. | ___ | ___ |
| **4.** Ensured that an equipment inventory was conducted and that discrepancies were reported before taking responsibility for the inventory. | ___ | ___ |
| **5.** Ensured that traffic control devices were employed according to each gate's special orders. | ___ | ___ |
| **6.** Inspected the condition of the facilities and surrounding areas. | ___ | ___ |
| **7.** Ensured that all gate discrepancies and facility deficiencies were reported or annotated for the record. | ___ | ___ |

| Performance Measures | GO | NO GO |
| --- | --- | --- |
| **8.** Reviewed the preliminary investigation reports generated on violations identified by access control personnel. | —— | —— |
| **9.** Ensured appropriate interaction by personnel with the public. | —— | —— |
| **10.** Ensured that immediate gate closure procedures were performed according to the appropriate FPCON level and the local SOP. | —— | —— |
| **11.** Ensured that a gate equipment inventory was conducted and that the facility was inspected before the end of the shift. | —— | —— |

**Evaluation Guidance:** Refer to chapter 1, paragraph 1-4 b (6).

**References**

**Required:** AR 190-13 and AR 525-13

**Related:** AR 190-14, AR 190-45, DA Form 2823, DA Form 3881, DA Form 3946, DA Form 3975, DA Form 4002, DA Form 4137, DD Form 1408, DD Form 1920, FM 3-19.13, FM 19-25, and STP 19-31B1-SM

**SUBJECT AREA 27: RISK MANAGEMENT**

## 153-001-3000

### Employ the CRM Process and Principles and Apply Them to Operations

**Conditions:** As a small unit leader, in a garrison or tactical environment, given a subordinate who has completed the composite risk management (CRM) process for a mission, activity, or task, access to FM 5-19 and a completed DA Form 7566 (*Composite Risk Management Worksheet*).

**Standards:** Verify that the CRM process has been conducted for a given mission, task, or activity by a subordinate leader.

**Performance Steps**

1. Verify that a risk assessment has been conducted.

   **a.** Ensure that mission, enemy, terrain and weather, troops and support available, time available, civil considerations (METT-TC) factors have been considered and hazards associated with the mission or task have been identified.

   **b.** Ensure that the level of risk for each hazard was correctly estimated.

2. Verify that controls have been developed and properly implemented.

   **a.** Ensure appropriate controls have been developed.

   **b.** Ensure that the residual risk for each hazard has been reassessed.

   **c.** Ensure that controls have been properly implemented. (How can we ensure controls are implemented?)

| Performance Steps |
|---|

**3.** Verify how the controls will be supervised and evaluated.

**a.** Verify the technique for monitoring the controls.

**b.** Verify how well the controls and the risk management process worked. (How can we verify that the process worked?)

**4.** Verify that the overall residual risk for the mission or task is correct.

**5.** Verify that the overall risk for the mission has been accepted by the proper level of the chain of command.

**Evaluation Preparation:** *Setup*: To evaluate this task, you will require a subordinate who has received a mission or task which may be in the form of an operations order (OPORD), fragmentary order (FRAGO), warning order, patrol order, training task, and so forth.

*Brief Soldier:* Tell the Soldier that you are going to evaluate him/her on his/her ability to supervise the implementation of the CRM process.

| Performance Measures | GO | NO GO |
|---|---|---|
| **1.** Verified that all of the hazards associated with the mission or task (a minimum of one, depending on the mission or task) had been identified. | —— | —— |
| **2.** Verified that the level of risk for each hazard had been correctly estimated. | —— | —— |
| **3.** Verified that the controls for each hazard were appropriate. | —— | —— |
| **4.** Verified the residual risk for each hazard. | —— | —— |
| **5.** Verified methods for implementing the controls. | —— | —— |
| **6.** Verified how controls would be supervised and evaluated. | —— | —— |
| **7.** Verified the overall residual risk for the mission, task, or activity. | —— | —— |
| **8.** Verified that the overall level of risk had been approved at the proper level of command. | —— | —— |

**Evaluation Guidance:** Refer to chapter 1, paragraph 1-4 b (6).

**References**

**Required:** DA Form 7566, FM 5-0, and FM 5-19

## SUBJECT AREA 3: CHEMICAL, BIOLOGICAL, RADIOLOGICAL, AND NUCLEAR

# 031-503-4002

### Radiological, Prepare a Unit for a Chemical, Biological, and Nuclear (CBRN) Attack

**Conditions:** You are given the commander's guidance, a radio, pioneer tools (ax, shovel, mattock), chemical alarms, chemical-agent detector kits, covering materials (plastic sheets, ponchos, tarpaulins), FM 3-11.4, and Soldiers who have their mission-oriented protective posture (MOPP) gear.

**Standards:** Prepare a unit for a CBRN attack by supervising unit preparation; preparing and protecting materiel from becoming damaged, contaminated, or inoperable, and preventing unit personnel from becoming casualties. The standard is not degraded if performed in MOPP4.

**Performance Steps**

**1.** Supervise unit preparation for a nuclear attack.

**a.** Individual protection.

**(1)** Ensure that personnel are sheltered in well-constructed fighting positions with overhead cover, bunkers, and/or armored vehicles.

**(2)** Ensure that personnel cover all exposed skin (roll sleeves down, button collars).

*Note*: A handkerchief or similar cloth may be worn over the nose and mouth to prevent inhaling contaminated dust.

**b.** Position.

**(1)** Ensure that the terrain is used effectively to minimize the effects of an attack.

**(2)** Ensure that the type of available shelter selected (covered foxholes, field-expedient overhead cover, buildings, tents, and armored vehicles) provides the best protection from weapon effects.

**c.** Materiel.

**(1)** Ensure that supplies, equipment, and vehicles are dispersed and/or dug in as much as possible. Ensure that explosives, ammunition, and flammables (fuel and oil) have been dispersed and/or dug in. Cover them if possible.

**(2)** Ensure that existing cover provides protection and natural shielding for vehicles, supplies, and equipment in the event of a nuclear explosion. Ensure that vans are parked so that their air conditioner intakes are opposite the prevailing wind direction. Ensure that air conditioners are turned off and intakes are covered with nonporous materials (plastic sheets or ponchos).

**(3)** Ensure that flammable debris is kept to a minimum. Ensure that small objects are secured to minimize the danger of casualties and damage from flying debris.

**Performance Steps**

**(4)** Ensure that all food and water are tightly sealed in containers and secured under available cover.

**(5)** Ensure that all electronic equipment and radios are turned off if not required for use.

**(6)** Ensure that power cables, antennas, and unused electronic equipment are disconnected and removed from power mounts.

**(7)** Ensure that communications and electronics equipment are placed inside bunkers or armored vehicles to enhance protection against an electromagnetic pulse (EMP).

**d.** Unit. Ensure that CBRN equipment operators prepare the equipment for use.

**2.** Supervise unit preparation for a biological attack.

**a.** Individual protection.

**(1)** Coordinate with medical personnel for needed immunizations.

**(2)** Ensure that Soldiers practice good hygiene and field sanitation procedures.

**(3)** Ensure that Soldiers are in good physical condition (well rested, well fed, and healthy).

**(4)** Ensure that Soldiers button clothing and cover exposed skin or wear the appropriate MOPP level.

**(5)** Ensure that Soldiers check protective masks and individual equipment for serviceability.

**(6)** Coordinate with supply personnel for needed parts and equipment.

**b.** Position.

**(1)** Ensure that the terrain is used effectively to minimize the effects of an attack.

**(2)** Ensure that the type of available shelter selected (covered foxholes, field-expedient overhead cover, buildings, tents, and armored vehicles) provides the best protection from weapon effects.

**c.** Materiel.

**(1)** Ensure that all food and water are tightly sealed in containers and secured under available cover.

**(2)** Ensure that unused supplies and equipment are covered to avoid surface contamination from an aerial spray attack.

**(3)** Ensure that vans are parked so that their air conditioner intakes are opposite the prevailing wind direction. Ensure that air conditioners are turned off and intakes are covered with nonporous materials (plastic sheets or ponchos).

**3.** Supervise unit preparation for a chemical attack.

**a.** Individual protection.

**(1)** Ensure that Soldiers are in good physical condition (well rested, well fed, and healthy).

| Performance Steps |
| --- |

**(2)** Ensure that Soldiers check protective equipment for serviceability and that they coordinate with supply personnel for replacement parts and equipment.

**(3)** Ensure that the appropriate MOPP level is assumed.

**(4)** Ensure that individuals seek available overhead cover or use ponchos, plastic sheets, tarpaulins, and so on to cover their positions.

**(5)** Ensure that antidotes and decontaminating kits are serviceable and readily available.

**b.** Position.

**(1)** Ensure that the terrain is used effectively to minimize the effects of an attack.

**(2)** Ensure that the type of available shelter selected (covered foxholes, field-expedient overhead cover, buildings, tents, and armored vehicles) provides the best protection from weapon effects.

**c.** Materiel.

**(1)** Ensure that all equipment, supplies, and vehicles are dispersed as much as possible. Ensure that these items are covered with nonporous materials (plastic sheets or ponchos). As a last resort, use dense foliage.

**(2)** Ensure that vans are parked so that their air conditioner intakes are opposite the prevailing wind direction. Ensure that air conditioners are turned off and intakes are covered with nonporous materials (plastic sheets or ponchos).

**(3)** Ensure that unpackaged food is in sealed containers. Ensure that Soldiers open food only when they are ready to eat it, keep water in sealed containers, and cover items if possible.

**d.** Unit. Ensure that operators of CBRN equipment prepare the equipment for use.

**Evaluation Preparation:** *Setup*: To evaluate this task, choose a site that allows the dispersing and/or digging in of vehicles, supplies, and equipment. You may decide to quiz the Soldier on performance measures that are difficult to evaluate otherwise, such as individual preparation for a biological attack.

*Brief Soldier:* Tell the Soldier that he/she will be evaluated on his/her ability to supervise unit preparations for a CBRN attack.

| Performance Measures | GO | NO GO |
| --- | --- | --- |
| **1.** Supervised unit preparation for a nuclear attack. | ____ | ____ |
| **2.** Supervised unit preparation for a biological attack. | ____ | ____ |
| **3.** Supervised unit preparation for a chemical attack. | ____ | ____ |

**Evaluation Guidance:** Refer to chapter 1, paragraph 1-4 b (6).

**References**

**Required:** FM 3-11.4

**Related:**

# 071-326-5775

## Coordinate with an Adjacent Platoon

**Conditions:** Given a platoon-sized element with two adjacent platoon-sized elements.

**Standards:** Coordinated with the adjacent platoon-sized elements for offensive and defensive operations; ensured mutually supporting positions, fires, and signals.

**Performance Steps**

**1.** Coordinate with adjacent elements. After receiving an order for an offensive or defensive operation and during your planning phase, you must consider coordination with adjacent elements. If you receive the order while all other platoon-sized element leaders are present, take that opportunity to coordinate as much as possible to avoid delays later in the operation. While many of the details that must be coordinated will vary with the situation, essential items must always be coordinated.

**2.** Coordinate in the offense. In the offense, you must coordinate—

**a.** Lateral distance between all attacking elements.

**b.** Movement routes, to ensure that mutual support by fire or maneuver can be maintained between the lead elements.

**c.** Visual signals such as arm-and-hand signals and pyrotechnics.

**d.** Radio call signs.

**3.** Coordinate in the defense. In the defense, you must coordinate to ensure that there are no gaps and that fires interlock and are mutually supporting. Information coordinated includes—

**a.** Location of positions (primary, alternate, and supplementary).

**b.** Dead space between units.

**c.** Locations of observation posts (OPs).

**d.** Signals.

**e.** Patrols and ambushes (size, type, time of departure and return, and routes).

**f.** Locations and types of obstacles.

**g.** Boundaries.

**h.** Control measures.

| Performance Steps |
|---|
| **4.** Allocate final protective fire. If a mortar or artillery final protective fire is allocated to the platoon-sized elements, it must be coordinated with the fire support team (FIST) forward observer (FO) and integrate it into the fire plan for the element. |

**Evaluation Preparation:** *Setup*: In the defense, provide a field location with varying terrain, two adjacent element leaders, and the last fighting position for each of the flanking elements. In the offensive, provide a field location with varying terrain and two element leaders from adjacent platoons.

*Brief Soldier:* As the center platoon leader, the Soldier must coordinate with both adjacent element leaders.

| Performance Measures | GO | NO GO |
|---|---|---|
| **1.** Coordinated with adjacent element leaders for offensive operations. | —— | —— |
| **a.** Lateral distance between attaching elements. | | |
| **b.** Movement routes. | | |
| **c.** Visual signals. | | |
| **d.** Radio call signs. | | |
| **e.** Boundaries. | | |
| **f.** Control measures. | | |
| **2.** Coordinated with adjacent leaders for defensive operations. | —— | —— |
| **a.** Location of primary position. | | |
| **b.** Location of alternate position. | | |
| **c.** Location of supplementary position. | | |
| **d.** Dead space between units. | | |
| **e.** Locations of OPs. | | |
| **f.** Signals. | | |
| **g.** Patrols and ambushes. | | |
| **h.** Locations and types of constructed obstacles. | | |
| **i.** Boundaries. | | |
| **j.** Control measures. | | |

**Evaluation Guidance:** Refer to chapter 1, paragraph 1-4 b (6).

**References**

**Required:** FM 3-21.8 (FM 7-8), FM 3-21.71, FM 3-90.1, and FM 7-7

**Related:**

## SUBJECT AREA 20: DEFENSE MEASURES

## 071-430-0006

### Conduct a Defense by a Platoon

**Conditions:** Given a specified area, a platoon with table of organization and equipment (TOE) equipment, and a requirement to defend that area.

**Standards:** Engage the threat according to the defensive plan, control fires, retain terrain, and destroy or repel the threat.

**Performance Steps**

**1.** Analyze the mission.

**a.** You are given a mission to defend by the company commander in an operation order (OPORD) or a warning order.

**b.** Identify both specified tasks and implied tasks.

**c.** Make an estimate of the situation using the factors of mission, enemy, terrain and weather, troops and support available, time available, civil considerations (METT-TC).

**2.** Issue a warning order to the squad leaders. The warning order should include the mission, time, place for issuing the OPORD, and any special instructions needed to start preparing for the mission.

**3.** Make a tentative plan.

**4.** Conduct a reconnaissance.

**a.** As a minimum, make a map reconnaissance.

**b.** If at all possible, conduct a ground reconnaissance.

**c.** Evaluate the terrain based on how the available observation, fields of fire, cover and concealment, obstacles, key terrain, and avenues of approach best support your mission and the commander's intent.

**d.** Request that the squad leaders, radio/telephone operator, and forward observers accompany you on the reconnaissance. You may need to take along some additional security depending on the tactical situation.

**Performance Steps**

**5.** Start necessary movement or preparations. The platoon sergeant should—

**a.** Initiate orders to prepare for any necessary movement and prepare for the defense as soon as the warning order has been issued.

**b.** Request and draw ammunition, rations, water, and any special equipment required.

**c.** Accomplish any needed maintenance.

**d.** Perform any movement required.

**e.** Accomplish this during the time you and your reconnaissance party are conducting the reconnaissance.

**6.** Complete the plan.

**a.** Complete the initial plan or revise it based on the continuing analysis and completion of the reconnaissance.

**b.** As a minimum, include the following in the plan:

**(1)** The deployment of squads.
**(2)** The deployment of key weapons.
**(3)** The use of indirect fire.
**(4)** The use of mines and obstacles.
**(5)** The establishment of security measures.
**(6)** The selection and operation of a command and observation posts.

**7.** Occupy the position.

**a.** Establish local security. Locate observation posts to make maximum use of long-range observations.

**b.** Position key weapons. Ensure that machine guns cover infantry avenues of approach, have primary and secondary sectors of fire and provide as much grazing fire as possible, and are assigned either a final protective line (FPL) or a principal direction of fire. Ensure that Javelins or Dragons cover armor avenues of approach, have primary and secondary sectors of fire, are positioned to engage targets from the flank, and are mutually supporting.

**c.** Clear fields of fire. Ensure that each fighting position clears its field of fire to engage the advancing enemy without exposing friendly positions. Range cards are prepared for each machine gun and Dragon.

**d.** Ensure that fighting positions have overlapping sectors of fire

**e.** Improve fighting positions as time becomes available. Improving overhead cover, aiming and limiting stakes, and camouflage are an ongoing activities.

**f.** Select and improve alternate and supplementary positions. Improve them as time becomes available.

| Performance Steps |
|---|

**8.** Emplace early warning devices. Use the platoon early warning system, if available. Set out trip flares. Use improvised early warning devices—such as noise makers, trip wire grenades, or other explosives.

**9.** Emplace hasty minefields and other obstacles. These should be covered by observation and fire.

**10.** Establish communication systems. Depend on radio as little as possible. Establish wire networks down to each squad and up to the company command post (CP).

**11.** Stockpile ammunition, water, food, and other supplies. Protect any materiel from direct fire and provide overhead protection.

**12.** Engage the enemy at maximum range. When the enemy appears in the platoon sector, engage the enemy with supporting direct and indirect fires. As the enemy comes within the range of your organic weapons, direct your gunners to start engaging the enemy.

**a.** When the enemy encounters your minefields and obstacles, use all friendly fires to breakup the enemy formations.

**b.** If the enemy is able to start an assault, repel or destroy the enemy by calling for final protective fires (FPF) of small arms, machine guns, mortars, and artillery.

**Evaluation Preparation:** *Setup*: Select an area in the field large enough for a platoon defensive position, including primary, alternate, and supplementary positions.

*Brief Soldier:* Tell the Soldier the platoon's sector of responsibility and that the Soldier is the acting platoon leader. Issue an OPORD (from the company commander) for a defense of the position.

| Performance Measures | GO | NO GO |
|---|---|---|
| **1.** Analyzed the mission. | —— | —— |
| **2.** Issued a warning order. | —— | —— |
| **3.** Made a tentative plan. | —— | —— |
| **4.** Conducted a reconnaissance. | —— | —— |
| **5.** Initiated necessary movement and preparation. | —— | —— |
| **6.** Completed the plan and issued the order. | —— | —— |
| **7.** Occupied the position. | —— | —— |
| **a.** Established local security. | | |
| **b.** Positioned key weapons. | | |

| Performance Measures | GO | NO GO |
|---|---|---|
| **c.** Ensured that fields of fire were cleared. | | |
| **d.** Ensured that sectors of fire overlapped. | | |
| **e.** Selected alternate and supplementary positions. | | |
| **8.** Ensured that early warning devices were installed. | —— | —— |
| **9.** Ensured that hasty protective minefields and other obstacles were constructed. | —— | —— |
| **10.** Established communication systems, wire, and radio. | —— | —— |
| **11.** Requested ammunition, food, water, and other supplies for stockpiling. | —— | —— |
| **12.** Engaged the enemy at maximum ranges. | —— | —— |
| **a.** Covered obstacles with fire. | | |
| **b.** Called for FPL and FPF. | | |

**Evaluation Guidance:** Refer to chapter 1, paragraph 1-4 b (6).

**References**

**Required:** FM 3-21.8 (FM 7-8), FM 3-21.71, FM 3-90.1, and FM 7-7

**Related:**

**SUBJECT AREA 22: UNIT OPERATIONS**

## 071-326-3013

### Conduct a Tactical Road March

**Conditions:** As an acting platoon sergeant/platoon leader, given a platoon with table of organization and equipment (TOE) weapons, personnel, equipment, operational vehicles with basic item issue (BII), a 1:50,000 military map of the road march area, a warning order, and an overlay of your route; a requirement to maintain radio listening silence until detected or engaged or until a spot report/status report (SPOTREP/STATREP) must be sent.

**Standards:** Within the time allowed in the warning order, conduct a tactical road march from one point to an assembly area; plan, organize, and control the road march and secure the assembly area.

**Performance Steps**

**1.** Control actions in the assembly area.

**a.** Ensure that the quartering party noncommissioned officer (NCO) guides the section/platoon into the area.

**b.** Clear the section/platoon release point quickly.

**Performance Steps**

**c.** Ensure that one man for each vehicle mans the crew-served weapon, monitors the radio, and observes the vehicle sector.

**d.** Set up the observation posts (OPs).

**e.** Check primary positions and sectors; adjust if necessary.

**f.** Camouflage positions.

**g.** Start the rest plan.

**h.** Ensure that the range cards are prepared.

**i.** Locate the troop command post.

**j.** Coordinate with flank elements.

**k.** Select and prepare alternate positions.

**l.** Prepare the platoon fire plan, brief the vehicle commanders, and send a copy of the fire plan to the troop commander.

**m.** Ensure that maintenance is performed.

**n.** Check personnel and equipment status; report to the team/troop commander.

**o.** Resupply.

**p.** Check security.

**2.** Plan the sequence for a tactical road march.

**a.** Prepare and issued the warning order.

**b.** Prepare an estimate of the situation.

**c.** Organize and dispatch reconnaissance and quartering parties according to the unit standing operating procedure (SOP).

**d.** Prepare detailed movement plans.

**(1)** Organize the march.
**(2)** Review reconnaissance information.

**e.** Prepare and issue the complete march order.

**3.** Issue the march order. The order includes the following:

**a.** Destination (map).

**b.** Route of march (map).

**c.** Location of start point (SP), critical points, and release point (map).

**d.** Start point time.

| Performance Steps |
|---|

**e.** March interval (in meters).

**f.** March speed.

**g.** Catch-up speed.

**h.** Time to leave present position.

**i.** Order of march.

**4.** Organize the march as provided in the following paragraphs.

**a.** March columns contain three elements: head, main body, and trail.

**(1)** The head is the first vehicle of the column and normally sets the pace.

**(2)** The main body is made up of the major elements of column serials and march units.

**(3)** The trail party follows the march column and includes personnel and equipment needed for emergency vehicle repair and recovery, medical aid and evaluation, and unscheduled refueling.

**b.** Ensure that the vehicle commanders assign sectors of observation so that there is 360-degree observation around their vehicles. Ensure that each vehicle commander designates an air guard to provide air security.

**5.** Control the march as provided below.

**a.** Follow the route specified in the warning order.

**b.** Enter the route of march at the SP.

**c.** Adjust the march at critical points if movement is slow or difficult or when vehicles lose their way.

**d.** Release sections to their mission responsibilities.

**e.** Ensure that the march and catch-up speeds are as specified in the warning order.

**f.** Maintain the order of march. This includes the order of squads in the column and the vehicles in each section.

**g.** Maintain the march interval.

**h.** Close the column intervals during periods of limited visibility.

**i.** Open the column during periods of good visibility.

**6.** Control actions at halts.

**a.** Establish security off road (when possible).

**b.** Man weapons and radio on each vehicle; ensure that each vehicle sector is observed.

| Performance Steps |
|---|

**c.** Maintain air guard.

**d.** Dismount OP if visibility is poor.

**e.** Post guides to help other traffic pass.

**f.** Determine the cause of unscheduled halts.

**g.** Eliminate the cause of the halt.

**h.** Report the status to the platoon leader/troop commander.

**i.** Ensure that maintenance is performed at scheduled halts.

**Evaluation Preparation:** *Setup*: At the test site, provide all equipment and materials given in the task conditions statement and the opportunity to issue a warning order. The task is performed along the specified route within the time allowed in the warning order. This task is performed as part of field training exercises.

*Brief Soldier:* Tell the Soldier to conduct a tactical road march using the equipment stated in the task conditions statement and the route specified in your warning order.

| Performance Measures | GO | NO GO |
|---|---|---|
| **1.** Conducted each step in the planning sequence for a tactical road march. | ___ | ___ |
| **2.** Ensured that all required actions in the assembly area were performed. | ___ | ___ |
| **3.** Issued the march order in its entirety. | ___ | ___ |
| **4.** Identified two of the three elements of the march columns. | ___ | ___ |
| **5.** Organized the march. | ___ | ___ |
| **6.** Ensured that all actions at halts were performed. | ___ | ___ |

**Evaluation Guidance:** Refer to chapter 1, paragraph 1-4 b (6).

**References**

**Required:** FM 3-21.8 (FM 7-8), FM 3-21.71, FM 7-7, and FM 17-95

**Related:**

## 071-720-0015
### Conduct an Area Reconnaissance by a Platoon

**Conditions:** Given a 1:50,000 map, a lensatic compass, and a mission to conduct an area reconnaissance within a specified time.

**Standards:** Satisfactorily perform the following within the time specified by the commander:

1. Organize the platoon into the command, reconnaissance, and security elements needed to accomplish the mission.
2. Conduct a reconnaissance using the surveillance or vantage-point method.
3. Obtain and report information about the terrain and enemy within the specified area.
4. Enter and leave the target area without being detected by the enemy.

**Performance Steps**

**1.** Estimate the situation. After receiving the reconnaissance mission, develop an estimate of the situation. Base the estimate on current intelligence about the enemy in the vicinity of the target area and on the capabilities of the unit. While planning for the mission, have the unit prepare for the mission also. Tailor the organization to best support the mission. The reconnaissance element of a platoon will normally be no larger than a squad.

**2.** Plan details. Develop the overall plan with a consideration of the following factors:

**a.** Use Intelligence. All reconnaissance operations must be based on the best information available as to actual conditions in the objective area.

**b.** Use deceptive measures. The success of reconnaissance operations is determined, to a large extent, on deception measures and on undetected infiltration and exfiltration.

**c.** Use the smallest unit possible to accomplish the mission. This decreases the possibility of enemy detection. Though only a small element reconnoiters, the parent unit must be large enough to provide security or support if the reconnaissance is detected or the element is engaged by an enemy force.

**d.** Remain undetected. The unit uses stealth, camouflage, concealment, and sound and light discipline. These techniques allow the unit to take advantage of periods of limited visibility to avoid contact and to get near, or on, the objective.

**e.** Use surveillance, target acquisition, and night observation (STANO) devices. The unit uses STANO devices to help it move and gain information about the enemy. Based on intelligence reports, consider the enemy's detection devices. When the enemy may have detection devices, use passive devices to decrease the probability of enemy detection.

**f.** Rehearse. After intelligence has been analyzed, the plan developed, special items of equipment procured and issued, and the troops briefed, the unit rehearses the plan. This rehearsal is a key factor that enhances the probability of the success of the operation. Rehearsals are as detailed as time will allow and include dry runs and briefings with repetition and questioning as needed to ensure understanding of the plan. Contingency plans are also rehearsed; these plans must be repeated by the reconnaissance members to ensure that they are understood.

**Performance Steps**

**g.** Minimize audio and electronic communications. Constraints on communication depend on enemy detection abilities and on how time sensitive the information obtained from the enemy is. There may be instances where the importance of the information may require an immediate report. The unit's existence could be threatened. Often, a one-time radio contact during the mission is necessary.

**h.** Inspect. The planning phase of the operation will include at least one inspection of all members of the reconnaissance force, to include their equipment. Only essential equipment, identified by the platoon leader as equipment required for mission accomplishment, will be carried. Special equipment must be closely inspected and safeguarded to ensure that it is functioning during the mission. Extra personnel and equipment assigned to the unit must also be carefully inspected and monitored prior to and during the operation. Any shortcomings found in personnel or equipment are corrected before the operation begins. Thorough inspections and supervision of personnel and equipment before the operation reduce the probabilities of compromise or failure.

**3.** Assign subordinate missions. Regardless of the types of reconnaissance, units are normally assigned one of three subordinate missions: Command and control, reconnaissance of the objective, or security of the force.

**a.** Command and control. The commander of the unit conducting a reconnaissance normally requires a small command group to assist in communicating with higher headquarters, subordinate elements, and supporting forces and to coordinate and control supporting elements, fire support, and air or water transport for the operation. For small operations, this group may consist of only the commander and a radio operator. For larger operations, the commander may require intelligence, logistics, and fire support elements, with adequate communication personnel for sustained 24-hour operations. The command group is always kept as small as possible.

**b.** Reconnaissance of the objective. The element with the reconnaissance mission approaches the target using stealth and concealment. All plans and applicable contingencies are conducted with the major effort made toward obtaining the information required while remaining undetected. The reconnaissance element must skillfully avoid all known and discovered enemy sensing devices; therefore, patience is important. Passive STANO devices will be used to observe activities at the objective. Information received about the target may be transmitted back to the appropriate headquarters by electronic means as it is observed. Or, the reconnaissance personnel may withdraw from the target and disseminate information by other means. The reconnaissance site should be sterilized before withdrawal. Withdrawal from the area must be as skillful, patient, and precise as the movement into it.

| Performance Steps |
| --- |

**c.** Security of the force. The reconnaissance element(s) with this mission must provide the commander with sufficient warning of the location and movement of enemy forces to permit the parent force to take evasive action or, when this is not possible, to provide covering fires that permit withdrawal of the reconnaissance element. Only if warnings are timely and information is accurate does the commander have the time and space to react. It is also the only way that, if the reconnaissance element is detected, the commander can arrange to give it sufficient overwatching suppressive fires or time to evade and withdraw safely.

**4.** Conduct an area reconnaissance and obtain information about a specific location and the area immediately around it (for example, road junctions, hills, bridges, enemy positions). Designate the location of the objective by either grid coordinates or a map overlay with a boundary line drawn around the area.

**a.** Once given an area reconnaissance mission, move the platoon to the appointed area in the shortest amount of time. This normally involves traveling along existing roads using the appropriate movement techniques. During this movement to an area, the platoon reports and bypasses enemy opposition unless ordered to do otherwise.

**b.** When the platoon reaches its area, it halts and sets up an objective rallying point (ORP). Once the ORP has been set up, the objective can be reconnoitered in one of two ways.

**(1)** When the terrain permits the security element to move to a position to overwatch the reconnaissance element, the leader may decide to have small reconnaissance teams move to each surveillance point or vantage point around the objective instead of having the entire element move as a unit from point to point. After the objective has been reconnoitered, the elements return to the ORP and information is issued. The patrol then returns to friendly lines.

**(2)** When the terrain does not allow the platoon to secure the objective area, the platoon leaves a security element at the ORP and uses reconnaissance and security (R&S) teams to reconnoiter the objective. These teams move to different surveillance points or vantage points, from which they reconnoiter the objective. Once the objective has been reconnoitered, the R&S team returns to the ORP, shares the information, and returns to friendly lines.

**Evaluation Preparation:** *Setup*: At the test site, provide all equipment and material given in the task conditions statement.

*Brief Soldier:* Tell the Soldier to organize and conduct an area reconnaissance using the surveillance or vantage-point method. Obtain and report any information about the terrain and enemy within the specified area. Tell the Soldier to complete the mission within the time specified by the commander. You will act as the battalion intelligence officer (S2) to provide answers to any questions the Soldier may ask.

| Performance Measures | GO | NO GO |
| --- | --- | --- |
| **1.** Established a plan of action based on the mission and the enemy situation. | —— | —— |
| **2.** Conducted a reconnaissance. | —— | —— |

| Performance Measures | GO | NO GO |
|---|---|---|
| **3.** Conducted deceptive measures during infiltration and exfiltration. | ___ | ___ |
| **4.** Used the smallest unit required to conduct the reconnaissance. | ___ | ___ |
| **5.** Applied stealth, camouflage, and concealment techniques along with noise and light discipline to avoid detection. | ___ | ___ |
| **6.** Conducted pre-mission inspection and rehearsal. | ___ | ___ |
| **7.** Used the correct size command and control group for the mission. | ___ | ___ |
| **8.** Used the correct movement techniques during the mission. | ___ | ___ |
| **9.** Used the correct security techniques during the mission. | ___ | ___ |

**Evaluation Guidance:** Refer to chapter 1, paragraph 1-4 b (6).

**References**

**Required:** FM 3-21.8 (FM 7-8), FM 3-21.71, and FM 7-7

**Related:**

## 091-CTT-4001

### Supervise Maintenance Operations

**Conditions:** In a contemporary operational environment, given a platoon,equipment, maintenance facility/site, standing operating procedures (SOPs), and applicable references.

**Standards:** Establish and maintain an effective platoon maintenance program according to SOPs and applicable references.

**Performance Steps**

**1.** Ensure that platoon maintenance procedures are according to the unit's SOP which is derived from AR 750-1, DA Pam 750-35, and local command policies.

**a.** Review the unit SOP for maintenance.

**b.** Inform unit personnel of changes in policy and new policy that impacts the unit SOP.

**2.** Provide maintenance guidance and assistance to a platoon.

**a.** Inspect platoon maintenance operations regularly.

**b.** Identify shortcomings.

| Performance Steps |
|---|

c. Make recommendations for corrective actions.

d. Provide training in maintenance procedures.

3. Provide maintenance management to a platoon.

a. Review the Non-Mission Capable Report.

b. Coordinate with field level maintenance for assistance and repairs above operator level.

c. Prioritize maintenance efforts of the platoon.

d. Direct cross-leveling of personnel, basic issue items (BII), and workload within a platoon.

| Performance Measures | GO | NO GO |
|---|---|---|
| **1.** Demonstrated knowledge of platoon maintenance procedures according to Army regulations and the unit SOP. | —— | —— |
| **2.** Stated correctly the correct period in which an inspection of unit operations should be conducted. | —— | —— |
| **3.** Identified correctly shortcomings in the platoon's maintenance program. | —— | —— |
| **4.** Chose the proper corrective action for deficiencies in the platoon maintenance program. | —— | —— |
| **5.** Determined correctly where to find resources for training a platoon on maintenance tasks. | —— | —— |
| **6.** Identified correctly the data fields on the Non-Mission Capable Report. | —— | —— |
| **7.** Identified inaccuracies of the Non-Mission Capable Report. | —— | —— |
| **8.** Identified the correct level of repair of an identified fault. | —— | —— |
| **9.** Identified correctly which piece of equipment in a platoon had maintenance priority. | —— | —— |
| **10.** Identified correctly workload issues in a platoon which could be solved by moving personnel or BII. | —— | —— |

**Evaluation Guidance:** Refer to chapter 1, paragraph 1-4 b (6).

**References**

**Required:** AR 385-10, AR 385-40, AR 385-55, AR 600-55, AR 700-4, AR 700-138, AR 710-2, AR 725-50, AR 735-5, AR 750-1, AR 750-43, DA Pam 25-30, DA Pam 710-2-1, DA Pam 710-2-2, DA Pam 750-1, DA Pam 750-8, DA Pam 750-35, and FM 4-30.3

**Related:**

**SUBJECT AREA 23: SECURITY AND CONTROL**

## 191-379-4408

### Plan Security for a Command Post

**Conditions:** You are a military police platoon sergeant and your platoon has been given a mission to provide security for a command post (CP). You will have orders; information on the size and layout of the CP (massed or dispersed); information on mission, enemy, terrain and weather, troops and support available, time available, civil considerations (METT-TC); and access to the provost marshal and headquarters personnel.

**Standards:** Develop a security plan that includes all the required elements to detect the enemy and to defend the CP before the enemy can move within direct-fire range.

**Performance Steps**

**1.** Determine the number of personnel available to conduct CP security, to include augmented personnel, such as corps assets and the division band. Consider METT-TC and the CP size and layout.

*Note*: For dispersed CPs, military police security is concentrated toward providing early warning through screening operations. For massed CPs, the type of military police security provided depends on the presence or absence of augmentation. Only when a CP is massed can the military police provide close-in security through augmentation from corps assets.

**2.** Identify the personnel to secure critical facilities within the CP.

*Note*: Critical facilities within the CP requiring security and controlled access include the tactical operations center, the war room, the communications center, the facilities for special intelligence, and the commander's quarters.

**3.** Conduct a reconnaissance of the routes to the CP and areas around it.

**4.** Plan the best method and/or mix of forces for security, varying between massed and dispersed CPs, to include static posts, traffic control posts (TCPs), listening posts (LPs), observation posts (OPs), access control, mobile patrols, and quick-response forces, both organic and augmenting.

**5.** Determine the number of personnel required to provide close-in security for the commander within the area of operations (AO).

**6.** Determine the amount of Class IX barrier materials needed.

**7.** Request the barrier materials through the appropriate channels.

| Performance Steps |
|---|
| **8.** Obtain an access list of personnel authorized in the CP from the provost marshal or headquarters personnel. |
| **9.** Prepare plans for the internal security of the CP. Coordinate directly with the existing security forces, access control personnel, headquarters personnel, and the provost marshal, as appropriate. |
| **10.** Plan for the security of tactical CPs when established and when in transit. |
| **11.** Prepare a standing operating procedure (SOP) and/or special orders detailing duties, responsibilities, and procedures to be used to provide security to the CP. |
| **12.** Brief all Soldiers included in the CP security on the tactical situation and the defensive plans. |

**Evaluation Preparation:** *Setup*: This task may be evaluated in the field or in a classroom environment. Provide the Soldier with information on the size and layout of the CP (massed or dispersed), a map of the area, and METT-TC. The evaluator will act as provost marshal and headquarters personnel and answer any questions the Soldier may have for which information has not been provided.

| Performance Measures | GO | NO GO |
|---|---|---|
| **1.** Determined the number of personnel available to conduct CP security, to include augmented personnel. | ___ | ___ |
| **2.** Identified personnel to secure critical facilities within the CP. | ___ | ___ |
| **3.** Conducted reconnaissance of the routes to the CP and the areas around it. | ___ | ___ |
| **4.** Planned the best method and/or mix of forces for security, varying between massed and dispersed CPs, to include static posts, TCPs, LPs, OPs, access control, mobile patrols, and quick-response forces, both organic and augmenting. | ___ | ___ |
| **5.** Determined the number of personnel required to provide close-in security for the commander within the AO. | ___ | ___ |
| **6.** Determined the amount of Class IX barrier materials needed. | ___ | ___ |
| **7.** Requested the barrier materials through the appropriate channels. | ___ | ___ |
| **8.** Obtained an access list of personnel authorized in the CP from the provost marshal or headquarters personnel. | ___ | ___ |

| Performance Measures | GO | NO GO |
|---|---|---|
| **9.** Prepared plans for the internal security of the CP. Coordinated directly with the existing security forces, access control personnel, headquarters personnel, and the provost marshal, as appropriate. | —— | —— |
| **10.** Planned for the security of tactical CPs when established and when in transit. | —— | —— |
| **11.** Prepared an SOP and/or special orders detailing duties, responsibilities, and procedures to be used to provide security to the CP. | —— | —— |
| **12.** Briefed all Soldiers included in the CP security on the tactical situation and the defensive plans. | —— | —— |

**Evaluation Guidance:** Refer to chapter 1, paragraph 1-4 b (6).

**References**

**Required:** FM 3-19.4

**Related:**

## 191-379-4440

### Supervise the Evacuation of Dislocated Civilians

**Conditions:** You are given a situation with dislocated civilians (DCs), a platoon of fully equipped Soldiers, and vehicles for movement (as appropriate).

**Standards:** Plan and supervise the evacuation of, accountability for, and welfare and safety of DCs.

**Performance Steps**

**1.** Analyze the DC evacuation movement order/situation.

- **a.** Provide a warning order to the platoon.
- **b.** Determine the number of DCs to be moved.
- **c.** Determine the time available and the time necessary for the mission.
- **d.** Identify the location for DCs to be evacuated to.
  - **(1)** Identify safety factors.
  - **(2)** Identify logistical support, if possible.
- **e.** Determine the required resources to complete the evacuation of DCs.
  - **(1)** Personnel.
  - **(2)** Movement type.
    - **(a)** Wheeled vehicles.
    - **(b)** Transports.

| Performance Steps |
| --- |

(c) Aircraft.

(d) Foot.

(3) Fuel.

(4) Water.

(5) Food.

(6) Medical.

**f.** Perform a route reconnaissance.

(1) Map reconnaissance.

(2) Actual reconnaissance (mission/time dependent).

**2.** Coordinate and prepare necessary resources to evacuate DCs.

**a.** Determine the number of Soldiers required for the mission.

**b.** Determine the number and type of vehicles required for the mission, if appropriate.

**c.** Locate necessary logistics from platoon or company resources.

(1) Water.

(2) Food.

(3) Emergency medical.

**d.** Coordinate the movement of DCs through higher headquarters.

(1) Determine the movement location assignment.

(2) Coordinate with the movement control center.

**e.** Process DCs.

(1) Prepare a log of DCs for movement.

(a) Identify DCs.

(b) Record DCs' passport number, identification card number, and so forth).

(c) Document DCs' location of origin.

(d) Document DCs' anticipated direction of movement.

(2) Screen DCs to determine their grouping.

(a) Group I.

(b) Group II.

(c) Group III.

**3.** Brief Soldiers on the DC evacuation mission.

**a.** Identify mission requirements.

**b.** Identify convoy methods.

**c.** Identify routes.

**Performance Steps**

**d.** Identify emergency response tactics for—

**(1)** Direct fire.

**(2)** Indirect fire.

**(3)** Air attack.

**(4)** Riot.

**e.** Identify secondary marshaling areas following attacks along established route.

**f.** Identify any known friendly fire support along identified routes.

**g.** Identify alternative routes in case of an emergency.

**h.** Identify communications.

**i.** Identify weapons.

**4.** Prepare DCs for movement.

**a.** Brief DCs (in their language when possible) on actions to be taken.

**b.** Address immediate human needs (water consumption, latrine use, and so forth) before movement.

**c.** Load vehicles, if appropriate.

***Note***: This movement is regarded as a low intensity/no evacuee threat type mission. No physical restraints should be used. Normal precautions (based on the threat; the local standing operating procedure; and the mission, enemy, terrain and weather, troops and support available, time available, civil considerations [METT-TC]) should be considered in this movement. Crew-served weapons should be manned and gunners should be seated where appropriate within the vehicles for convoy security.

**5.** Evacuate DCs along an approved main supply route (MSR) or an approved movement route (assigned by the movement control center).

**6.** Secure the DC evacuation location/site.

**7.** Move DCs to the evacuation location/site.

**8.** Notify higher headquarters of mission completion.

**a.** Report the number of DCs moved.

**b.** Report the location of DCs.

**c.** Report issues important to the movement.

**(1)** Mishaps.

**(2)** Injuries.

**(3)** Deaths.

**(4)** Platoon status.

Evaluation Preparation: *Setup*: Provide the Soldier with the materials and equipment listed in the conditions statement.

*Brief Soldier:* Tell the Soldier to perform the necessary requirements to successfully evacuate DCs to an evacuation location/site.

| Performance Measures | GO | NO GO |
|---|---|---|
| **1.** Analyzed the DC evacuation movement order/situation. | —— | —— |
| **2.** Coordinated and prepared necessary resources to evacuate DCs. | —— | —— |
| **3.** Briefed Soldiers on the DC evacuation mission. | —— | —— |
| **4.** Prepared DCs for movement. | —— | —— |
| **5.** Evacuated DCs along an approved MSR or an approved movement route. | —— | —— |
| **6.** Secured the DC evacuation location/site. | —— | —— |
| **7.** Moved DCs to the evacuation location/site. | —— | —— |
| **8.** Notified higher headquarters of mission completion. | —— | —— |

**Evaluation Guidance:** Refer to chapter 1, paragraph 1-4 b (6).

**References**

**Required:**

**Related:** FM 3-19.4

## 191-379-5403

### Supervise a Riot/Crowd Control Operation with a Platoon-Sized Element

**Conditions:** You are given a riot or crowd control situation and a platoon of Soldiers fully equipped with riot control gear, FM 3-19.15, and STP 19-31B1-SM.

**Standards:** Based on the situation, select the best formations to control or disperse the crowd. Correctly form the platoon into the formations. Move the platoon through the formations to disperse or move the crowd, and reassemble the platoon from the formations after the crowd has been controlled or dispersed.

| Performance Steps |
|---|

**1.** Analyze mission requirements.

- **a.** Determine the size of the area the platoon is to maneuver through.
- **b.** Determine the approximate number of people in the crowd.
- **c.** Determine the goals of the mission as follows:
  - **(1)** Move the crowd.
  - **(2)** Apprehend individuals, as appropriate.
  - **(3)** Extract individuals, as appropriate.
- **d.** Complete a risk assessment.

**Performance Steps**

**e.** Determine the necessary mission support requirements (if any). Determine if there is a need for—

**(1)** Backup support.

**(2)** Vehicle support.

**(3)** Photographer support (video preferred).

**(4)** Communications support.

**(5)** Medical support.

**f.** Identify any potential threats to the rear of the formation once any crowd control movements have been completed.

**g.** Determine the platoon or operation exit strategy.

**2.** Determine the type of formations to be used based on the mission.

***Note***: The platoon leader determines into which formation to assemble the platoon. The formation to be used depends on the structure and temperament of the crowd. A squad, platoon, or larger unit may be employed and crowd control formations can be adapted to fit the unit's organization. For more information on formations, see FM 3-19.15.

**a.** Form the platoon into a platoon line formation with one squad in direct (figure 191-379-5403-1).

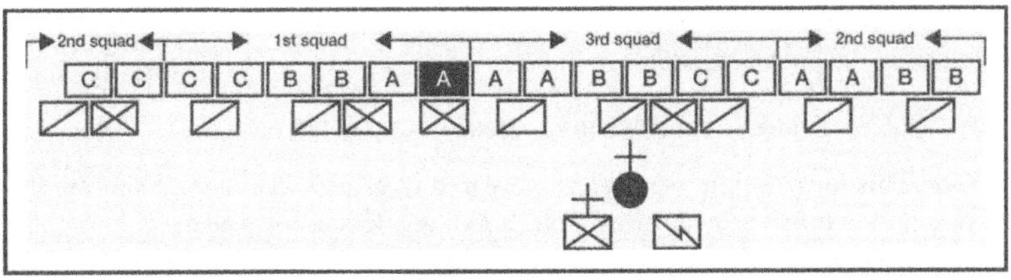


**Figure 191-379-5403-1. Platoon line formation**

***Note***: This formation is used to push a crowd in a straight line backwards.

Performance Steps

**b.** Form the platoon into a platoon echelon left (or right) formation (figure 191-379-5403-2).

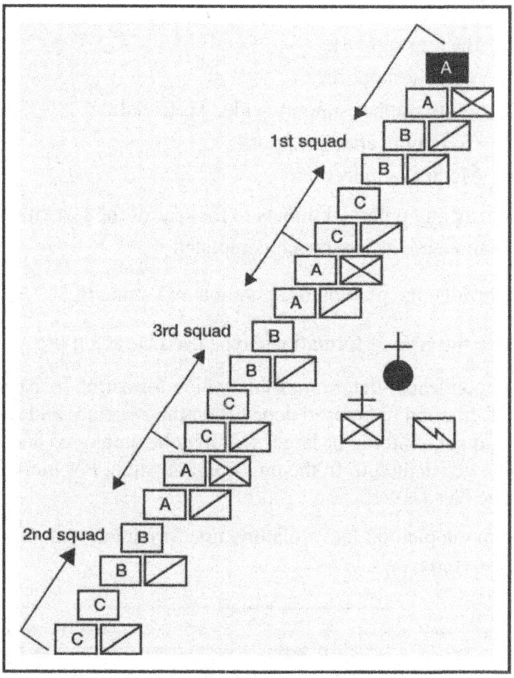


**Figure 191-379-5403-2. Platoon echelon left formation**

*Note*: This formation is used offensively to turn groups in either open or built-up areas and to move crowds away from buildings, fences, or walls.

Performance Steps

**c.** Form the platoon into a platoon echelon left (or right) formation with one support squad in general support, the same as with the platoon line formation (figure 191-379-5403-3).

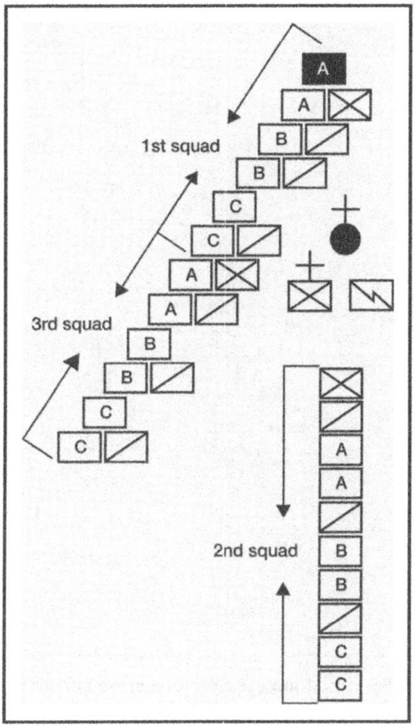


**Figure 191-379-5403-3. Platoon echelon left with one support squad in general support**

*Note*: This formation is used offensively to turn groups in either open or built-up areas and to move crowds away from buildings, fences, or walls.

Performance Steps

**d.** Form the platoon into a platoon echelon left (or right) formation with one support squad in lateral support, the same as with the platoon line formation (figure 191-379-5403-4).

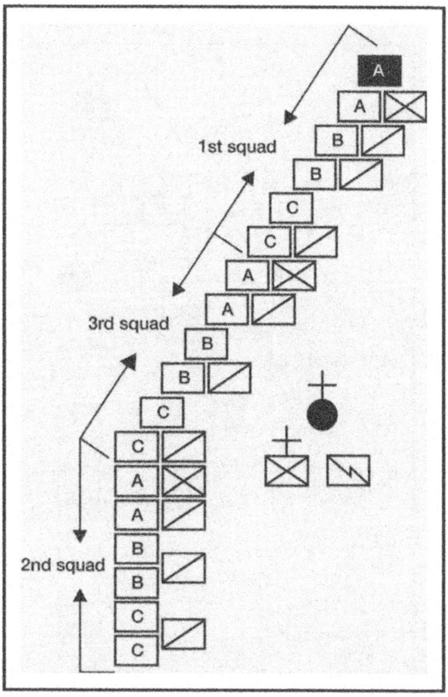


**Figure 191-379-5403-4. Platoon echelon formation left with one squad in lateral support**

***Note***: This formation is used offensively to turn groups in either open or built-up areas and to move crowds away from buildings, fences, or walls.

**e.** Form the platoon into a platoon echelon left (or right) formation with one support squad in direct support, the same as with the platoon line formation (figure 191-379-5403-5).

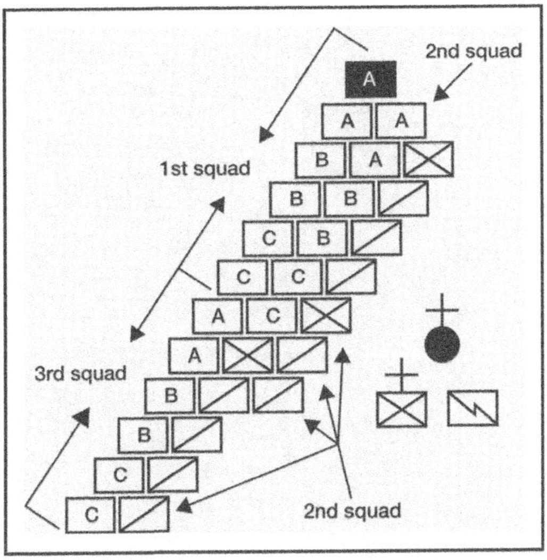


**Figure 191-379-5403-5. Platoon echelon formation left with one squad in direct support**

*Note*: This formation is used offensively to turn groups in either open or built-up areas and to move crowds away from buildings, fences, or walls.

**f.** Form the platoon into a platoon wedge formation (figure 191-379-5403-6).

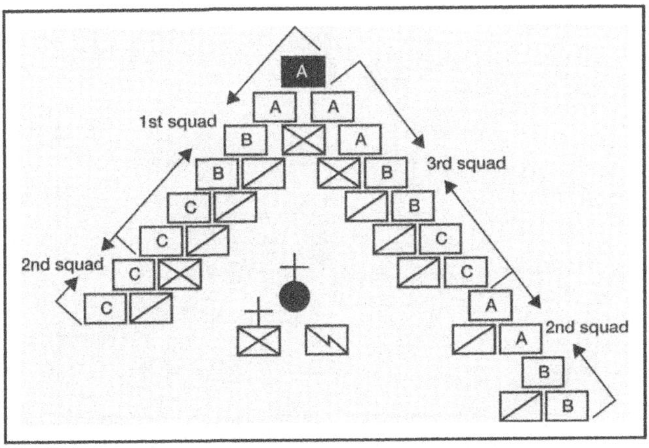


**Figure 191-379-5403-6. Platoon wedge formation**

*Note*: This formation is used offensively to penetrate or split crowds.

**g.** Form the platoon into a platoon wedge formation with one squad in general support (figure 191-379-5403-7).

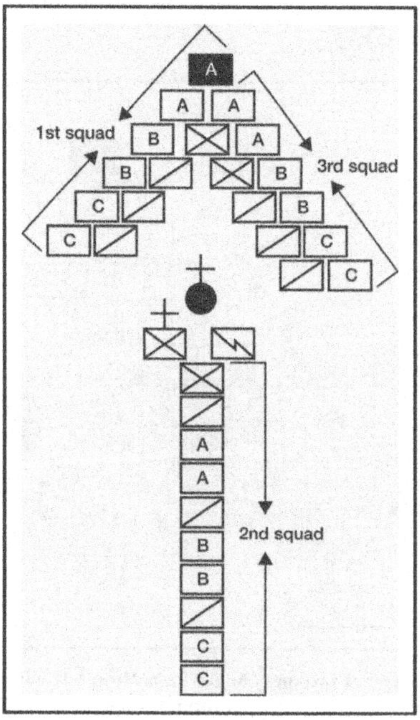


**Figure 191-379-5403-7. Platoon wedge formation with one squad in general support**

*Note*: This formation is used offensively to penetrate or split crowds.

**h.** Form the platoon into a platoon wedge formation with one squad in lateral support (figure 191-379-5403-8).

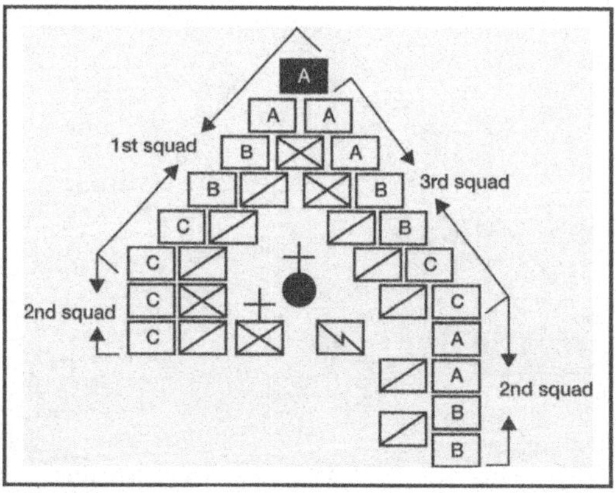


**Figure 191-379-5403-8. Platoon wedge formation with one squad in lateral support**

*Note*: This formation is used offensively to penetrate or split crowds.

**Performance Steps**

**i.** Form the platoon into a platoon wedge formation with one squad in direct support (figure 191-379-5403-9).

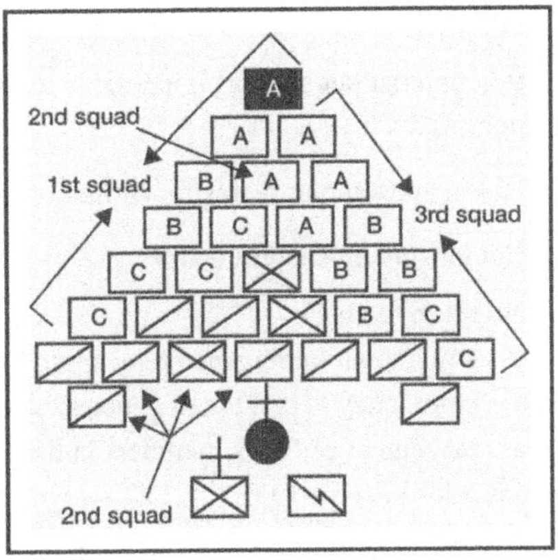


**Figure 191-379-5403-9. Platoon wedge formation with one squad in direct support**

**3.** Identify any nonlethal weapons (NLWs) requirement and assign platoon members to NLW positions, as appropriate:

**a.** NLW capability SE Kit.

**b.** 12-gauge shotguns.

**c.** M203 grenade launcher.

***Note***: If nonlethal munitions are to be used, they should be addressed within the rules of engagement (ROE) and disseminated to the lowest level. This requires that all personnel have a clear understanding of the ROE and the leader's intent.

**4.** Identify any lethal overwatch team requirements.

***Note***: Lethal overwatch teams are assigned based on the mission, enemy, terrain and weather, troops and support available, time available, civil considerations (METT-TC); mission risk factors; availability; and the environment in which the mission is to be performed. For example, a hostile fire zone would require the most support from lethal overwatch teams. Examples of when lethal overwatch teams are authorized to employ lethal force are situations involving—

- Self-defense and the defense of others based on imminent danger of death.
- Assets of national security that are in danger.
- Assets inherently dangerous, such as explosives or chemical or nuclear materials.

**a.** Brief lethal overwatch teams on the ROE, and ensure that they maintain communication with the movement leader.

**Performance Steps**

**b.** Ensure that sniper cover support is available to the formation movement, where appropriate.

**c.** Ensure that vehicle crew support is available to the formation movement, where appropriate.

**d.** Ensure that countersniper support is available to the formation movement, where appropriate.

**5.** Organize the platoon and support teams for formation and action.

**a.** Brief the platoon and all support teams:

**(1)** Discuss mission goals.
**(2)** Discuss any known intelligence relevant to the mission.
**(3)** Discuss emergency action contingencies.
**(4)** Discuss placement of NLW members in the platoon formation.
**(5)** Discuss rules of force/ROE.

**b.** Direct all support teams to take their positions.

**6.** Move the platoon from the staging area to the mission area.

**7.** Form the platoon into a platoon line formation.

**a.** Give the preparatory command, "Platoon on line," while simultaneously using hand-and-arm signals. (Raise both arms straight out to the side, arms and hands extended and palms down.)

**b.** Give the execution command, "Move," and point to the place where you want the formation.

**c.** Ensure that the squad leaders of the 1st and 2d squads give the command, "Follow me," while the leader of the 3d squad gives the command, "Stand fast." The rest of the 2d squad forms a line to the right of the baseman. The 1st squad forms a line to the left.

**d.** Ensure that the squad leader of the 3d squad gives the command, "Follow me," after the 1st and 2d squads have assumed their positions.

**e.** Ensure that the 3d squad forms a line to the right of the 2d squad.

***Note***: All squads dress on the 2d squad.

**8.** Form the platoon into a platoon line formation with one squad in general support.

**a.** Form the platoon line with one squad in general support in a column of twos.

**(1)** Give the command, "Platoon line, 2d squad in support in a column of twos."

**(2)** Ensure that the leaders of the 1st and 3d squads command, "Follow me," and the leader of the 2d squad commands, "Stand fast."

| Performance Steps |
| --- |

**(3)** Give the command, "Move," and point to the place where you want the formation. The 1st and 3d squads move forward to their appointed places. The 2d squad, at the command of its squad leader, executes a column of twos to the right.

**b.** Form the platoon into a platoon line with one squad in support in a single column. The support squad, usually the 2d squad, may also be formed into a single column. The command is, "Platoon line, 2d squad in support, move." The 1st and 3d squads execute the formation, while the 2d squad remains in a column formation.

**9.** Form the platoon into a platoon line formation with one squad in lateral support.

**a.** Give the command, "Platoon line, 2d squad in lateral support."

**b.** Give the command, "Move," and point to the place where you want the formation.

**c.** Ensure that after the 1st squad forms a line to the left of the baseman, the 3d squad forms a line to the right of the baseman. After the line has been formed, the squad leader for the 2d squad commands, "Odd-numbered Soldiers, follow me," and the assistant squad leader commands, "Even-numbered Soldiers, follow me." The odd-numbered Soldiers move out to the left and form a column behind the last Soldier in the 3d squad. The even-numbered Soldiers move out to the right and form a column behind the last Soldier in the 1st squad.

**10.** Form the platoon into a platoon line formation with one squad in direct support.

**a.** Give the command, "Platoon line, 2d squad in direct support." The 1st and 3d squads will execute a line.

**b.** Ensure that the leaders of the 2d squad command, "Stand fast."

**c.** Ensure that after the 1st and 3d squads move to the appointed place and execute the line, the 2d squad executes a similar line to the rear of the leading line. The 2d squad forms to the right of its baseman. The 2d squad can be shifted to the right or left to support any segment of the formation.

***Note***: Ensure that the 2d squad closes and covers the intervals between the elements in the leading line.

**11.** Form the platoon into a platoon echelon left (or right) formation.

**a.** Give the command, "Platoon echelon, left (or right)."

**b.** Give the command, "Move," and point to where you want the formation.

**Performance Steps**

**c.** Give hand-and-arm signals with voice commands. Extend one arm 45 degrees above and the other 45 degrees below the horizon, arms and hands extended. (When facing the platoon, the upper arm indicates the direction of the echelon.)

**d.** Ensure that the 1st squad leader commands, "Follow me," and the squad executes an echelon left.

**e.** Ensure that the 2d and 3d squad leaders command, "Stand fast."

**f.** Ensure that as each squad clears the column, the next squad moves to extend the echelon formed by the preceding squad.

***Note***: For an echelon right, the 3d squad executes an echelon right with the rest of the squads following.

**12.** Form the platoon into a platoon echelon left (or right) with one support squad in general support, the same as with the platoon line formation.

**a.** Give the command, "Platoon echelon, left (or right), 2d squad in general support."

**b.** Give the command, "Move," and point to the place where you want the formation.

**c.** Give hand-and-arm signals with voice commands. Extend one arm 45 degrees above and the other 45 degrees below the horizon, arms and hands extended. (When facing the platoon, the upper arm indicates the direction of the echelon.)

**d.** Ensure that the 1st squad leader commands, "Follow me," and the squad executes an echelon left (or right).

**e.** Ensure that the 2d and 3d squad leaders command, "Stand fast."

**f.** Ensure that as the 1st squad clears the column, the 3d squad moves to extend the echelon formed by the 1st squad. The 2d squad, at the command of its squad leader, executes a column of twos to the left (or right).

**13.** Form the platoon into a platoon echelon left (or right) with one support squad in lateral support, the same as with the platoon line formation.

**a.** Give the command, "Platoon echelon, left (or right), 2d squad in lateral support."

**b.** Give the command, "Move," and point to the place where you want the formation.

**c.** Give hand-and-arm signals with voice commands. Extend one arm 45 degrees above and the other 45 degrees below the horizon, arms and hands extended. (When facing the platoon, the upper arm indicates the direction of the echelon.)

**Performance Steps**

**d.** Ensure that the 1st squad leader commands, "Follow me," and the squad executes an echelon left (or right).

**e.** Ensure that the 2d and 3d squad leaders command, "Stand fast."

**f.** Ensure that as the 1st squad clears the column, the 3d squad moves to extend the echelon formed by the 1st squad. After the echelon has formed, the squad leader for the 2d squad commands, "Odd-numbered Soldiers, follow me," and the assistant squad leader commands, "Even-numbered Soldiers, follow me." The odd-numbered Soldiers move out to the left (or right) and form a column behind the last Soldier in the 3d squad. The even-numbered Soldiers move out to the left (or right) and form a column behind the first Soldier in the 1st squad.

**14.** Form the platoon into a platoon echelon left (or right) formation with one support squad in direct support, the same as with the platoon line formation.

**a.** Give the command, "Platoon echelon, left (or right), 2d squad in direct support."

**b.** Give the command, "Move," and point to the place where you want the formation.

**c.** Give hand-and-arm signals with voice commands. Extend one arm 45 degrees above and the other 45 degrees below the horizon, arms and hands extended. (When facing the platoon, the upper arm indicates the direction of the echelon.)

**d.** Ensure that the 1st squad leader commands, "Follow me," and the squad executes an echelon left (or right).

**e.** Ensure that the 2d and 3d squad leaders command, "Stand fast."

**f.** Ensure that as the 1st squad clears the column, the 3d squad moves to extend the echelon formed by the 1st squad. After the echelon has formed, the 2d squad executes a similar echelon to the rear of the leading echelon. The 2d squad will form to the left (or right) of its baseman. The 2d squad can be shifted to the left or right to support any segment of the formation.

**15.** Form the platoon in a platoon wedge formation.

**a.** Give the command, "Platoon wedge," followed by the command, "Move." Give hand-and-arm signals with voice commands. Extend both arms down to the sides at 45 degrees below the horizon, palms down and toward the body. Then point to where you want the formation.

**b.** Ensure that the 1st and 3d squad leaders command, "Follow me," while the 2d squad leader commands, "Stand fast."

**Performance Steps**

**c.** Ensure that the 1st and 3d squads move to the front and, when the last element of the squads have cleared the front of the 2d squad, the 2d squad leader commands, "Follow me." The squad moves to the left and right respectively.

**d.** Ensure that the 3d squad executes an echelon right off the baseman and the 1st squad executes an echelon left.

**e.** Ensure that the odd-numbered Soldiers of the 2d squad form an echelon left on the rear element of the 1st squad and the even-numbered Soldiers form an echelon right on the rear element of the 3d squad.

**16.** Form the platoon in a platoon wedge formation with one squad in general support in a column of twos.

**a.** Give the command, "Platoon wedge, 2d squad in support in a column of twos, move."

**b.** Ensure that the 1st and 3d squads execute the wedge and that the 2d squad, at the command of its squad leader, executes a column of twos to the right.

***Note***: The support squad, usually the 2d squad, may also be formed into a single column. The command for this would be, "Platoon wedge, 2d squad in support, move." The 1st and 3d squads execute the formation, while the 2d squad remains in a column formation.

**17.** Form the platoon into a platoon wedge formation with one squad in lateral support.

**a.** Give the command, "Platoon wedge, 2d squad in lateral support, move." The 1st and 3d squads execute the wedge, while the 2d squad stands fast.

**b.** Ensure that after the wedge has been formed, the squad leader for the 2d squad commands, "Odd-numbered Soldiers, follow me," and the assistant squad leader commands, "Even-numbered Soldiers, follow me." The odd-numbered Soldiers move out to the left and form a column behind the last Soldier in the 3d squad. The even-numbered Soldiers move out to the right and form a column behind the last Soldier in the 1st squad.

**c.** Give the command, "2d squad, lateral support, move," to move the 2d squad from general support to lateral support.

**d.** Give the command, "2d squad, extend the wedge, move," to have the 2d squad join the wedge from either general support or lateral support. The 2d squad leader and the assistant squad leader command, "Follow me," and move out to the left and right, respectively, to extend the wedge on the 1st and 3d squads.

**Performance Steps**

**18.** Form the platoon into a platoon wedge formation with one squad in direct support.

**a.** Give the command, "Platoon wedge, 2d squad in direct support, move."

**b.** Ensure that the 1st and 3d squads execute a wedge. The 2d squad executes a squad wedge behind, and centered on, the leading wedge.

**c.** Ensure that the Soldiers in the supporting wedge cover the intervals between the Soldiers in the leading wedge.

**19.** Move the formation using the command, "Forward, march."

***Note***: This will begin the forward movement of the formation. Left and right formation movement adjustments will be made by using "Echelon left or echelon right." Use other movement commands, such as mark time or halt, as appropriate.

**a.** Determine the flow of the crowd as your team advances, and make commands to the team for formation movements to compensate for the crowds movement. Ensure that the squad achieves the goals assigned.

***Note***: Squad leaders and squad members should be mindful of safety factors in formation movements, such as airborne projectiles, active antagonistic members in the crowd that may need removal, possible incoming direct fire, and peripheral view checks of left and right squad members.

**b.** Give the command, "Halt," once movement of the crowd is complete according to mission goals.

**c.** Review the crowd dispersal and reevaluate it to ensure that the mission is complete.

**d.** Check peripherals (side and rear) to ensure security and safety of the squad.

**e.** Communicate your intentions with any support teams to remove your squad before reassemble.

**20.** Reassemble the platoon from a formation.

**a.** Move to the rear of the platoon where you want the platoon members to assemble and give the command, "Platoon assemble, move," while making a circular motion above your head with your right hand.

**b.** Ensure that the baseman of each squad does an about face and that all other squad members do a facing movement toward their baseman

**c.** Ensure that the platoon comes to port arms upon hearing the command, "Platoon assemble."

| Performance Steps |
|---|
| **d.** Ensure that upon the command, "Move," the 2d squad leader commands, "Follow me." The 1st and 3d squad leaders command, "Stand fast." |
| **e.** Ensure that as the 2d squad clears the formation, the 1st and 3d squad leaders command, "Follow me." The 1st and 3d squads follow, moving at double time, and dress to the right of the 3d squad until the platoon is fully formed. |

**Evaluation Preparation:** *Setup*: Provide the Soldier with the materials and equipment listed in the conditions statement.

*Brief Soldier:* Tell the Soldier that this task may be performed to standard without regards to sequence of formations and that all formations in the task must be performed.

| Performance Measures | GO | NO GO |
|---|---|---|
| **1.** Analyzed mission requirements. | —— | —— |
| **2.** Determined the type of formations to be used based on the mission. | —— | —— |
| **3.** Identified any NLWs requirements. | —— | —— |
| **4.** Identified any lethal overwatch team requirements. | —— | —— |
| **5.** Organized the platoon and support teams for formation and action. | —— | —— |
| **6.** Moved the platoon from the staging area to the mission area. | —— | —— |
| **7.** Formed the platoon into a platoon line formation. | —— | —— |
| **8.** Formed the platoon into a platoon line formation with one squad in general support. | —— | —— |
| **9.** Formed the platoon into a platoon line formation with one squad in lateral support. | —— | —— |
| **10.** Formed the platoon into a platoon line formation with one squad in direct support. | —— | —— |
| **11.** Formed the platoon into a platoon echelon left (or right) formation. | —— | —— |
| **12.** Formed the platoon into a platoon echelon left (or right) formation with one support squad in general support. | —— | —— |
| **13.** Formed the platoon into a platoon echelon left (or right) formation with one support squad in lateral support. | —— | —— |
| **14.** Formed the platoon into a platoon echelon left (or right) formation with one support squad in direct support. | —— | —— |
| **15.** Formed the platoon into a platoon wedge formation. | —— | —— |

| Performance Measures | GO | NO GO |
|---|---|---|
| **16.** Formed the platoon into a platoon wedge formation with one squad in general support in a column of twos. | ___ | ___ |
| **17.** Formed the platoon into a platoon wedge formation with one squad in lateral support. | ___ | ___ |
| **18.** Formed the platoon into a platoon wedge formation with one squad in direct support. | ___ | ___ |
| **19.** Moved the formation using the command, "Forward, march". | ___ | ___ |
| **20.** Reassembled the platoon from a formation. | ___ | ___ |

**Evaluation Guidance:** Refer to chapter 1, paragraph 1-4 b (6).

**References**

**Required:** STP 19-31B1-SM

**Related:**

**SUBJECT AREA 26: CRIME PREVENTION**

## 191-379-4425

### Implement the Unit's Crime Prevention Program

**Conditions:** You are given the unit crime prevention standing operating procedures (SOP), the results of the last organizational inspection program (OIP), a copy of the installation crime prevention program, and FM 3-19.30.

**Standards:** Implement the unit crime prevention program according to the unit SOP and the installation crime prevention program.

**Performance Steps**

**1.** Review the unit crime prevention SOP and the installation crime prevention program to identify unique unit/installation needs and requirements.

**2.** Review the results of the last unit OIP for deficiencies identified during the last inspection.

**3.** Conduct crime prevention surveys according to the unit crime prevention SOP.

**4.** Ensure that unit equipment is marked according to Army regulations and command guidance.

***Note***: Equipment markings vary greatly depending on the type of equipment assigned to the unit. Check the unit SOPs, technical manuals, and/or Army regulations to determine the correct procedures for marking unit equipment.

| Performance Steps |
| --- |

**5.** Brief unit Soldiers on "Operation ID" to include the following:

- **a.** Methods of identifying personal property.
- **b.** Recording personal property
- **c.** Identification numbering.
- **d.** Standard Army numbering system.

**6.** Attend the battalion or brigade crime prevention working groups as required by the SOP and/or the installation crime prevention program.

**7.** Conduct crime prevention briefings to unit personnel, as required by the unit SOP and/or the installation crime prevention program.

| Performance Measures | GO | NO GO |
| --- | --- | --- |
| **1.** Reviewed the unit crime prevention SOP and the installation crime prevention program to identify unique unit/installation needs and requirements. | ___ | ___ |
| **2.** Reviewed the results of the last unit OIP for deficiencies identified during the last inspection. | ___ | ___ |
| **3.** Conducted crime prevention surveys according to the unit crime prevention SOP. | ___ | ___ |
| **4.** Ensured that unit equipment was marked according to Army regulations and command guidance. | ___ | ___ |
| **5.** Briefed unit Soldiers on "Operation ID." | ___ | ___ |
| **6.** Attended the battalion or brigade crime prevention working groups, as required by the SOP and/or the installation crime prevention program. | ___ | ___ |
| **7.** Conducted crime prevention briefings to unit personnel, as required by the unit SOP and/or the installation crime prevention program. | ___ | ___ |

**Evaluation Guidance:** Refer to chapter 1, paragraph 1-4 b (6).

**References**

**Required:** FM 3-19.30

**Related:**

## SUBJECT AREA 27: RISK MANAGEMENT

# 153-001-4000

**Integrate Risk Management into Mission Plans**

**Conditions:** As a company level senior leader, given a mission or task, in a garrison or tactical environment, access to FM 5-19, and a DA Form 7566 (*Composite Risk Management Worksheet*).

**Standards:** Demonstrate how the five steps of composite risk management (CRM) apply to the eight steps of the troop-leading procedures (TLPs).

**Performance Steps**

1. Describe the relationship between the five steps of CRM and the eight steps of TLP. See figure 153-001-4000-1.

| TROOP LEADING PROCEDURES | RISK MANAGEMENT STEPS: STEP 1 Identify Hazard | STEP 2 Assess Hazard | STEP 3 Develop Controls (+) Risk Decision | STEP 4 Implement Controls | STEP 5 Supervise (+) Evaluate |
|---|---|---|---|---|---|
| Receive Mission | X | X | | | |
| Issue Warning Order | X | X | X | | |
| Make a Tentative Plan | X | X | X | | |
| Initiate Movement | X | X | X | X | |
| Conduct Recon | X | X | X | X | |
| Complete Plan | X | X | X | X | |
| Issue Order | | | X | X | X |
| Supervise & Refine | | | | X | X |

**Figure 153-001-4000-1. TLP and CRM relationship**

2. Identify risk guidance received as part of the assigned mission order.

**Evaluation Preparation:** *Setup*: To evaluate this task, you need an assigned mission or task which may be in the form of an operations order (OPORD), fragmentary order (FRAGO), warning order, patrol order, training task, and so forth. The Soldier will apply the CRM process to the assigned mission or task as an integrated part of TLPs.

*Brief Soldier:* Tell the Soldier that you are going to evaluate him/her on his/her ability to apply the CRM process to TLPs.

| Performance Measures | GO | NO GO |
|---|---|---|
| **1.** Described the relationship between the five steps of CRM and the eight steps of TLP. | ___ | ___ |
| **2.** Identified the risk guidance contained in the mission order. | ___ | ___ |

**Evaluation Guidance:** Refer to chapter 1, paragraph 1-4 b (6).

**References**

**Required:** FM 5-0, FM 5-19, and DA Form 7566

**Related:**

**SUBJECT AREA 28: ADMINISTRATION/MANAGEMENT**

## 805C-PAD-2472

**Prepare a Duty Roster**

**Conditions:** You are a squad/section leader. Given requirements to provide Soldiers to perform additional duties on a recurring basis, DA Form 6 (*Duty Roster*) and AR 220-45.

**Standards:** Prepare a DA Form 6 for each duty requirement, list all eligible Soldiers, post only days on which a detail is required.

**Performance Steps**

**1.** Determine the requirements of the duty.

- **a.** Determine the period of the duty.
- **b.** Determine the nature of the duty performed.

**2.** Determine the eligible personnel.

- **a.** Determine grade of the individuals eligible for duty.
- **b.** Determine the special skills needed.

**3.** Prepare DA Form 6.

- **a.** Prepare a separate form for each recurring duty.
- **b.** Complete the administrative data.
  - **(1)** Enter the duty title.
  - **(2)** Enter the start date.
  - **(3)** Enter the month/day performed.
- **c.** Enter eligible Soldiers to perform duty.
- **d.** Post numbers for day detail is required.

**4.** Select Soldier(s) for duty

- **a.** Notify individual.
- **b.** Post duty roster.

**Evaluation Preparation:** *Setup:* To evaluate this task, give the Soldier a scenario that would require the individual to prepare a duty roster, post numbers, and select the individual to perform the duty. Provide the Soldier with a section roster, DA Form 6, pencil, and AR 220-45.

*Brief Soldier:* Tell the Soldier that he/she will be evaluated on his/her ability to prepare a duty roster.

| Performance Measures | GO | NO GO |
|---|---|---|
| **1.** Determined the requirements of the duty. | ___ | ___ |
| **a.** Determined the period of the duty. | | |
| **b.** Determined the nature of the duty performed. | | |
| **2.** Determined the eligible personnel. | ___ | ___ |
| **a.** Determined the grade of individuals eligible for duty. | | |
| **b.** Determined the special skills needed. | | |
| **3.** Prepared DA Form 6. | ___ | ___ |
| **a.** Prepared a separate form for each recurring duty. | | |
| **b.** Completed the administrative data. | | |
| **(1)** Entered the duty title. | | |
| **(2)** Entered the start date. | | |
| **(3)** Entered the month/day performed. | | |
| **c.** Entered the Soldiers eligible to perform duty. | | |
| **d.** Posted the numbers. | | |
| **4.** Selected the Soldier(s) for duty. | ___ | ___ |
| **a.** Notified the individual concerned. | | |
| **b.** Posted the duty roster. | | |

**Evaluation Guidance:** Refer to chapter 1, paragraph 1-4 b (6).

**References**

**Required:** AR 220-45 and DA Form 6

**Related:**

---

# 805C-PAD-4359

## Manage Soldier's Deployment Requirements

**Conditions:** Given a requirement to prepare a section for deployment, access to AR 600-8-101, local standing operating procedures (SOPs), standard office supplies, equipment, and assigned personnel.

**Standards:** Perform initial deployment activities to include verification of section roster, recall/alert roster, individual Soldier Readiness Processing (SRP) packets.

| Performance Steps |
|---|
| **1.** Review SRP packets to identify deployable and nondeployable Soldiers. |
| **2.** Review the section's recall procedures. |
| **3.** Identify all equipment the section needs, based on the mission, climate, location, and local SOP. |
| **4.** Identify all supplies the section needs, based on the mission, climate, location, and local SOP. |
| **5.** Identify any special items of clothing the section needs, based on the mission, climate, location, or local SOP. |

**Evaluation Preparation:** *Setup*: This task can be performed in an administrative environment. Soldiers should be informed to bring the section personnel roster, recall roster, and SRP packets. At the test site, verify that these files and documents have been properly prepared and are accurate and current.

*Soldier Brief*: Inform the Soldier that he/she is being evaluated on his/her ability to manage Soldier deployment requirements.

| Performance Measures | GO | NO GO |
|---|---|---|
| **1.** Reviewed SRP packets to identify deployable and nondeployable Soldiers. | —— | —— |
| **2.** Reviewed the section's recall procedures. | —— | —— |
| **3.** Identified all equipment the section needs, based on the mission, climate, location, and local SOP. | —— | —— |
| **4.** Identified all supplies the section needs, based on the mission, climate, location, and local SOP. | —— | —— |
| **5.** Identify any special items of clothing the section needs, based on the mission, climate, location, or local SOP. | —— | —— |

**Evaluation Guidance:** Refer to chapter 1, paragraph 1-4 b (6).

**References**

**Required:** AR 600-8-1 and SOP

**Related:**

## 805C-PAD-4550

### Prepare a Standing Operating Procedure (SOP)

**Conditions:** Given a requirement to prepare an SOP to standardize routine or recurring actions or functions, appropriate operational references or local policy guidance, and access to AR 25-50 and AR 380-5, computer with authorized software, and office supplies.

**Standards:** Prepare an SOP in the proper format and identify the purpose, distribution, and references. Apply the proper security classification marks (if applicable) and properly authenticate.

| Performance Steps |
|---|

1. Determine purpose.
   a. Determine requirement for action/function.
   b. Analyze intended outcome
2. Determine audience/distribution.
   a. Identify level of application.
   b. Identify units, section, and personnel involved.
3. Determine references for content.
4. Select content for the SOP.
5. Prepare the SOP in the format prescribed by local procedures.
6. Determine security classification requirements.
7. Apply the proper security classification marking.
8. Determine coordination requirement for approval.
9. Distribute the SOP for coordination.
10. Review coordination comments.
11. Prepare the final SOP.
    a. Incorporate appropriate comments.
    b. Coordinate final review.
12. Obtain the authentication.
13. Determine reproduction requirements.
14. Distribute the SOP as required by local procedures.

**Evaluation Preparation:** *Setup*: To evaluate this task, gather the items listed in the conditions statement. Have computer and printer or pencil and paper for Soldiers to prepare the SOP.

*Brief Soldier:* Give the Soldier a scenario that would provide all information necessary to perform the measures to include topic references.

| Performance Measures | GO | NO GO |
|---|---|---|
| **1.** Determined purpose. | ___ | ___ |
| **a.** Determined requirement for action/function. | | |
| **b.** Analyzed intended outcome. | | |
| **2.** Determined the audience/distribution. | ___ | ___ |
| **a.** Identified the level of application. | | |
| **b.** Identified units, section, and personnel involved. | | |

| Performance Measures | GO | NO GO |
|---|---|---|
| **3.** Determined references for content. | ___ | ___ |
| **4.** Selected content for the SOP. | ___ | ___ |
| **5.** Prepared the SOP in the format prescribed by local procedures. | ___ | ___ |
| **6.** Determined the security classification marking. | ___ | ___ |
| **7.** Applied the proper security classification marking. | ___ | ___ |
| **8.** Determined coordination requirement for approval. | ___ | ___ |
| **9.** Distributed the SOP for coordination. | ___ | ___ |
| **10.** Reviewed coordination comments. | ___ | ___ |
| **11.** Prepared the final SOP. | ___ | ___ |
| **a.** Incorporated appropriate comments. | | |
| **b.** Coordinated the final review. | | |
| **12.** Obtained the authentication. | ___ | ___ |
| **13.** Determined reproduction requirement. | ___ | ___ |
| **14.** Distributed the SOP as required by local procedures. | ___ | ___ |

**Evaluation Guidance:** Refer to chapter 1, paragraph 1-4 b (6).

**References**

**Required:** AR 25-50, AR 380-5, FM 101-5, and Software-Office

**Related:**

# 805C-PAD-4597

## Integrate Newly Assigned Soldiers

**Conditions:** Given a section/platoon, newly assigned Soldiers, and local policy.

**Standards:** Establish personnel accountability. Provide Soldiers with health and welfare items, initial area orientations, and special orders (if required). Verify Soldiers' qualifications, assign Soldiers to duty positions, inform the commander, and conduct initial counseling.

| Performance Steps |
|---|

**1.** Receive new Soldiers.

**a.** Provide health and welfare items.

**b.** Provide initial required area orientations and special orders.

**2.** Verify Soldiers' qualifications and equipment.

| Performance Steps |
| --- |
| **3.** Determine duty assignments. |
| **a.** Notify the commander. |
| **b.** Notify the Soldiers. |
| **4.** Provide initial counseling. |
| **5.** Update personnel accountability. |

**Evaluation Preparation:** *Setup*: Provide the Soldier with a scenario and sufficient information to accomplish the performance steps. Provide the Soldier with a section/platoon roster which indicates military occupational specialty (MOS), grades and filled/vacant positions, paper and pencils. Scenario should provide local requirements for personnel accountability, health and welfare items, and area orientations. Have the Soldier list the actions/items he/she would take or provide to integrate newly assigned Soldiers.

*Brief Soldier*: Inform the Soldier that he/she will be tested on his/her ability to receive and integrate new Soldiers into the section/platoon. Tell the Soldier he/she will list the actions/items necessary to properly integrate newly assigned Soldiers.

| Performance Measures | GO | NO GO |
| --- | --- | --- |
| **1.** Received new Soldiers. | —— | —— |
| **a.** Provided health and welfare items. | | |
| **b.** Provided initial area orientations and special orders. | | |
| **2.** Verified the Soldiers' qualifications and equipment. | —— | —— |
| **3.** Determined duty assignments. | —— | —— |
| **a.** Notified the commander. | | |
| **b.** Notified the Soldiers. | | |
| **4.** Provided initial counseling. | —— | —— |
| **5.** Updated personnel accountability. | —— | —— |

**Evaluation Guidance:** Refer to chapter 1, paragraph 1-4 b (6).

**References**

**Required:**

**Related:**

Appendix A

# Proponent School or Agency Codes

The first three digits of the task number identify the proponent school or agency responsible for the task. Record any comments or questions regarding the task summaries contained in this manual on a DA Form 2028 (*Recommended Changes to Publications and Blank Forms)* and send it to the proponent school with an information copy to—

Commander, U.S. Army Training Support Center
ATTN: ATIC-ITSC-CM
Fort Eustis, VA 23604-5166.

**Table A-1. Proponent School or Agency Codes**

| *School Code* | *Command* |
|---|---|
| MANSCEN<br>CM **031** | Commandant, U.S. Army Chemical, Biological, Radiological and Nuclear (CBRN) School<br>ATTN: ATSN-TI<br>464 MANSCEN Loop, Suite 2617<br>Fort Leonard Wood, MO 65473-8929 |
| MANSCEN<br>EN **052** | Commandant, U.S. Army Engineer School<br>ATTN: ATSE-DT (Individual Training Division)<br>320 MANSCEN Loop, Suite 370<br>Fort Leonard Wood, MO 65473 |
| MANSCEN<br>MP **191** | Commandant, United States Army Military Police School<br>ATTN: ATSJ-TD<br>401 MANSCEN Loop, Suite 2617<br>Fort Leonard Wood, MO 65473-8926 |
| FA<br>**061** | Directorate of Training and Doctrine<br>U.S. Army Field Artillery School<br>ATTN: ATSF-D<br>Fort Sill, OK 73503-5000 |
| IN<br>**071** | Commandant, U.S. Army Infantry School<br>ATTN: ATSH-OTT<br>Fort Benning, GA 31905-5593 |
| AHS<br>**081** | Department of Training Support<br>ATTN: MCCS-HTI<br>1750 Greeley Rd, Ste 135<br>Fort Sam Houston, TX 78234-5078 |
| CASCOM<br>**091** | U.S. Army Combined Arms Support Command (CASCOM) Training Directorate<br>USACASCOM, ATTN: ATCL-A<br>401 First St., Suite 227<br>Fort Lee, VA 23801-1511 |

| Table A-1. Proponent School or Agency Codes | |
|---|---|
| ***School Code*** | ***Command*** |
| TRADOC<br>Safety Office<br>**153** | TRADOC Command Safety Office<br>1 Bernard Road, Bldg 84<br>ATTN: ATCS-S<br>Fort Monroe, VA 23651-1048 |
| CAL<br>**158** | Director, Center for Army Leadership<br>ATTN: ATZL-CL<br>250 Gibbon Ave, Bldg 120 (Eisenhower Hall)<br>Fort Leavenworth, KS 66027 |
| JAG<br>**181** | Commandant, Judge Advocate General Legal Center<br>and School<br>ATTN: JAGS-TDD<br>600 Massie Road<br>Charlottesville, VA 22903-1781 |
| APAC<br>**224** | Director, Army Public Affairs Center<br>6 ACR Road, Bldg 8607<br>ATTN: SAPA-PA<br>Fort Meade, MD 20755-5650 |
| MI<br>**301** | Commander, USA Intelligence Center & Fort Huachuca<br>550 Cibeque Street, Suite 168<br>ATTN: ATZS-TDS-I<br>Fort Huachuca, AZ 85613-7002 |
| CASCOM<br>**551** | U.S. Army Combined Arms Support Command (CASCOM)<br>Training Directorate<br>USACASCOM, ATTN: ATCL-TDM<br>401 First St., Rm 101A<br>Fort Lee, VA 23801-1511 |
| SSI<br>**805C** | Commander, USA Soldier Support Institute<br>ATTN: ATSG-TDD<br>Fort Jackson, SC 29207-7065 |

Appendix B

# GUIDE TO FORMS

This appendix contains a list of forms pertinent to SMCT and CTT evaluations and administration. In the electronic, online version of this manual, hot links will, where indicated, provide downloadable and reproducible copies of the forms. The user can also visit the CTT site in the Reimer Digital Library where access to the ***CTT Manual*** and ***CTT Notice*** (Bulletin) provide additional helpful information and links. Some of these forms may be available in your unit.

**DA Form 5164-R** (*Hands On Evaluation*), used by CTT scorer to record GO/NO GO for each performance measure in a given task. Refer to the CTT Manual for the current FY (in the Reimer Digital Library) for links to DA Forms 5164-R that are overprinted with the performance measures for each task on the current CTT.

**CTT Performance Sheet** used by the CTT scorer to compile all GO/NO GO scores to produce an overall GO or NO GO for the Soldier's common task test. As with the Hands On Evaluation form above, the CTT Performance Sheet is different for each fiscal year; it is overprinted with the tasks from the current CTT and linked from the CTT Manual (in the Reimer Digital Library) for the current fiscal year.

**DA Form 5165-R** (*Field Expedient Squad Book*), used to record task proficiency for groups of Soldiers in a squad. This squad book should be especially beneficial for recording training results gathered during MTP, field exercises, or individual training sessions. Once training is completed, trainers may transfer information from DA Form 5165-R to the leader book (FM 3-21.10 [FM 7-10]).

**CTT Roll Up for SL 1-2 and SL 3-4**, used by trainers and administrators to record performances for larger, unit-sized groups of Soldiers. This form allows you to see and report the overall ratio of Soldiers passing the CTT to the total of those tested.

**DA Form 2028** (*Recommended Changes to Publications and Blank Forms*), used by Soldiers and trainers to record any comments or questions regarding the task summaries contained in this manual. Use the task proponent address for the appropriate proponent code given in appendix A (first three digits of the task number).

# Appendix C

# FUTURE CHANGES

## NONCOMMISSIONED OFFICER EDUCATION TRANSFORMATION

The transformation of the Basic Noncommissioned Officers Course (BNCOC) and the Advanced Noncommissioned Officers Course (ANCOC) to the Advanced Leaders Course (ALC) and Senior Leaders Course (SLC) with an implementation date of 1 October 2009. Current combat operations have demonstrated that Noncommissioned Officer Development must change to meet the operational Army needs and ensure relevance to the COE and future operations. This is more than just a name change, it is a downward migration of tasks to create the attributes and competencies required as the basis of the creation the broadly-skilled warrior.

The Advanced Leaders Course (ALC), formerly BNCOC will provide the Army's Junior Noncommissioned Officers the skills they need to conduct military operations, operations other than war, and logistics operations at the squad and platoon levels. Enrollment remains available to soldiers selected for promotion to Staff Sergeant.

The Senior Leaders Course (ALC), formerly ANCOC, will prepare the Senior Noncommissioned Officer to lead both platoon and company operations while additionally providing the necessary skills and attributes to lead and train Soldiers as a First Sergeant. Enrollment remains available to soldiers selected for promotion to Sergeant First Class.

There will be no change to current promotion policy. Current courses length will be maintained during migration of BNCOC to ALC and ANCOC to SLC. This transformation is part of an Army initiative to implement a life-long learning strategy for the Noncommissioned Officer

# Glossary

| | |
|---|---|
| **AA** | assembly area |
| **AAR** | after-action review |
| **AC** | alternating current |
| **ACE** | armored combat earthmover |
| **AIDS** | acquired immune deficiency syndrome |
| **AO** | area of operations |
| **APOD** | aerial port of debarkation |
| **ARNG** | Army National Guard |
| **AT** | antitank |
| **AUEL** | automated unit equipment list |
| **BAS** | battalion aid station |
| **BBPCT** | blocking, bracing, packing, crating, and tiedown |
| **BFV** | Bradley fighting vehicle |
| **BII** | basic issue items |
| **CAM** | chemical-agent monitor |
| **CASEVAC** | casualty evacuation |
| **CB** | chemical-biological |
| **CBRN** | chemical, biological, radiological, and nuclear |
| **CDR** | commander |
| **CED** | critical event debriefing |
| **CEE** | captured enemy equipment |
| **CEM** | captured enemy material |
| **CHRRS** | Community Housing Referral and Relocation Services |
| **CID** | Criminal Investigation Command |
| **CINC** | commander in chief |
| **COE** | contemporary operational environment |
| **COMMZ** | communications zone |
| **COMSEC** | communications security |
| **COSR** | combat and operational stress reaction |
| **CP** | command post |
| **CRM** | composite risk management |
| **DC** | dislocated civilian |
| **DCP** | detainee collection point |
| **DEL** | deployment equipment list |
| **DHA** | detainee holding area |
| **DL** | delay line |
| **DNBI** | disease and nonbattle injury |
| **DOD** | Department of Defense |
| **DRMO** | Defense Reutilization and Marketing Office |

| | |
|---|---|
| **DTG** | date-time group |
| **DU** | depleted uranium |
| **DULLRAM** | depleted uranium and low level radiation material |
| **EMP** | electromagnetic pulse |
| **EO** | equal opportunity |
| **EOA** | equal opportunity advisor |
| **EOAP** | equal opportunity action plan |
| **EOL** | equal opportunity leader |
| **EPW** | enemy prisoner of war |
| **ETA** | estimated time of arrival |
| **FDC** | fire direction center |
| **FEBA** | forward edge of the battle area |
| **FIST** | fire support team |
| **FO** | forward observer |
| **FPCON** | force protection condition |
| **FPF** | final protective fire |
| **FPL** | final protective line |
| **FRAGO** | fragmentary order |
| **FSMC** | forward support medical company |
| **FST** | field sanitation team |
| **FTX** | field training exercise |
| **GENTEXT** | general text |
| **GSA** | General Services Administration |
| **GTA** | graphic training aid |
| **HAZMAT** | hazardous material |
| **HEPA** | high-efficiency particulate air |
| **HN** | host nation |
| **HQDA** | Headquarters, Department of the Army |
| **HTD** | highway traffic division |
| **HW** | hazardous waste |
| **ID** | identification |
| **IDP** | initial delay position/individual development plan |
| **IED** | improvised explosive device |
| **IG** | inspector general |
| **IPR** | in-process review |
| **ISN** | internment serial number |
| **ISN** | internment serial number |
| **LC** | line of contact |
| **LD** | line of departure |
| **LLR** | low level radiation |
| **LP** | listening post |

| | |
|---|---|
| **MBA** | main battle area |
| **MCM** | maintenance control manual |
| **MDMP** | military decisionmaking process |
| **MEDEVAC** | medical evacuation |
| **MEDPLT** | medical platoon |
| **METL** | mission essential task list |
| **METT-TC** | mission, enemy, terrain and weather, troops and support available, time available, civil considerations |
| **MOPP** | mission-oriented protective posture |
| **MOS** | military occupational specialty |
| **MOUT** | military operations in urban terrain |
| **MSL** | mean sea level |
| **MSR** | main supply route |
| **MTP** | mission training plan |
| **NATO** | North Atlantic Treaty Organization |
| **NBC** | nuclear, biological, and chemical |
| **NCO** | noncommissioned officer |
| **NCOER** | noncommissioned officer evaluation report |
| **NCOIC** | noncommissioned officer in charge |
| **NLW** | nonlethal weapon |
| **NMC** | not mission capable |
| **NTAT** | not to accompany troops |
| **OAKOC** | observation and fields of fire, avenues of approach, key terrain, obstacles, and cover and concealment |
| **OEG** | operational exposure guidance |
| **OEL** | organization equipment list |
| **OIP** | organizational inspection program |
| **OP** | observation post |
| **OPLAN** | operations plan |
| **OPORD** | operation order |
| **OPSEC** | operations security |
| **ORP** | objective rallying point |
| **OSUT** | one-station unit training |
| **OT** | observer-target |
| **PAO** | public affairs office |
| **PATI** | protection assessment test instrument |
| **PATS** | Protection Assessment Test System |
| **PD** | point of departure |
| **PDF** | principal direction of fire |
| **PL** | phase line |
| **PM** | preventive medicine/provost marshal |
| **PMCS** | preventive maintenance checks and services |

| | |
|---|---|
| **PMM** | preventive medicine measure |
| **POD** | port of debarkation |
| **POL** | petroleum, oils, and lubricants |
| **POSH** | prevention of sexual harassment |
| **POW** | prisoner of war |
| **PP** | passage point |
| **PTSD** | post-traumatic stress disorder |
| **PVTMED** | preventive medicine |
| **R&S** | reconnaissance and security |
| **RB** | release other than attack biological |
| **RC** | release other than attack chemical |
| **RN** | release other than attack nuclear |
| **ROE** | rules of engagement |
| **ROTA** | release other than attack |
| **RP** | release point |
| **RU** | release other than attack unknown |
| **RUF** | rules for the use of force |
| **S1** | personnel staff officer |
| **S2** | brigade intelligence staff officer |
| **S3** | battalion or brigade operations staff officer |
| **SA** | semiannually |
| **SALUTE** | size, activity, location, unit, time, and equipment |
| **SIB** | separate infantry brigade |
| **SJA** | staff judge advocate |
| **SL** | skill level |
| **SM** | Soldier's manual |
| **SMCT** | Soldier's Manual of Common Tasks |
| **SOI** | signal operating instruction |
| **SOI** | signal operation instruction |
| **SOP** | standing operating procedure |
| **SP** | start point |
| **SPOD** | seaport of debarkation |
| **SPOTREP** | spot report |
| **SR** | supply route |
| **SRP** | Soldier Readiness Processing |
| **SSN** | social security number |
| **STANAG** | standardization agreements |
| **STANO** | surveillance, target acquisition, and night observation |
| **STATREP** | status report |
| **STP** | Soldier's training publication |
| **TAT** | to accompany troops |

| | |
|---|---|
| **TC** | training circular |
| **TC ACCIS** | Transportation Coordinators' Automated Command and Control Information System |
| **TC-AIMS II** | Transportation Coordinators' Automated Information for Movements System |
| **TCCC** | tactical combat casualty care |
| **TCP** | traffic control post |
| **TCP** | traffic control post, tactical control point |
| **TECHDOC** | technical documentation |
| **TECHINT** | technical intelligence |
| **TF** | task force |
| **TIF** | theater internment facility |
| **TIM** | toxic industrial material |
| **TLP** | troop-leading procedures |
| **TM** | team |
| **TM** | technical manual |
| **TOE** | table of organization and equipment |
| **TOW** | tube launched, optically tracked, wire guided |
| **TRP** | target reference point |
| **TSOP** | tactical standing operating procedure |
| **U.S.** | United States |
| **UCMJ** | Uniform Code of Military Justice |
| **UDL** | unit deployment list |
| **USAR** | United States Army Reserve |
| **WARNORD** | warning order |
| **WTBD** | warrior tasks and battle drills |
| **www** | worldwide web |

# References

## REQUIRED PUBLICATIONS

Required publications are sources that users must read to understand or comply with this publication.


**ARMY REGULATIONS**

AR 25-50. *Preparing and Managing Correspondence*. 3 June 2002.

AR 27-1. *Legal Services, Judge Advocate Legal Services*. 30 September 1996.

AR 27-10. *Military Justice*. 16 November 2005.

AR 55-162. *Permits for Oversize, Overweight, or Other Special Military Movements on Public Highways in the United States*. 1 January 1979

AR 165-1, *Chaplain Activities in the United States Army*. 25 March 2004.

AR 190-13. *The Army Physical Security Program*. 30 September 1993.

AR 220-45. *Duty Rosters*. 15 November 1975.

AR 360-1. *The Army Public Affairs* Program. 15 September 2000.

AR 380-5. *Department of the Army Information Security Program*. 29 September 2000.

AR 385-10. *The Army Safety Program*. 29 February 2000.

AR 385-40. *Accident Reporting and Records*. 1 November 1994.

AR 385-55. *Prevention of Motor Vehicle Accidents*. 12 March 1987.

AR 385-63. *Range Safety*. 19 May 2003.

AR 525-13. *Antiterrorism*. 4 January 2002.

AR 600-8-1. *Army Casualty Program*. 7 April 2006

AR 600-8-2. *Suspension of Favorable Personnel Actions (Flags)*. 23 December 2004.

AR 600-8-22. *Military Awards*. 11 December 2006.

AR 600-13. *Army Policy for the Assignment of Female Soldiers*. 27 March 1992.

AR 600-20. *Army Command Policy*. 18 April 2008.

AR 600-55. *The Army Driver and Operator Standardization Program (Selection, Training, Testing, and Licensing*). 31 December 1993.

AR 601-280. *Army Retention Program*. 31 January 2006.

AR 623-3. *Evaluation Reporting System*. 10 August 2007.

AR 630-10. *Absence Without Leave, Desertion, and Administration of Personnel*. 13 January 2006.

AR 635-200. *Active Duty Enlisted Administrative Separations*. 6 June 2005.

AR 670-1. *Wear and Appearance of Army Uniforms and Insignia*. 3 February 2005.

AR 700-4. *Logistics Assistance*. 17 March 2006.

AR 700-138. *Army Logistics Readiness and Sustainability*. 26 February 2004.

AR 710-2. *Supply Policy Below the National Level*. 8 July 2005.

AR 725-50. *Requisition, Receipt, and Issue System*. 15 November 1995.

AR 735-5. *Policies and Procedures for Property Accountability*. 28 February 2005.

AR 750-1. *Army Materiel Maintenance Policy*. 5 September 2006.

AR 750-43. *Army Test, Measurement and Diagnostic Equipment Program.* 14 December 2004.

**DEPARTMENT OF THE ARMY PAMPHLETS**

DA Pam 25-30. *Consolidated Index of Army Publications and Blank Forms.* 1 October 2006.

DA Pam 27-1. *Treaties Governing Land Warfare.* 7 December 1956.

DA Pam 385-64. *Ammunition and Explosives Safety Standards.* 15 December 1999.

DA Pam 600-67. *Effective Writing for Army Leaders.* 6 February 1986.

DA Pam 623-3. *Evaluation Reporting System.* 13 August 2007.

DA Pam 710-2-1. *Using Unit Supply System (Manual Procedures) (Standalone Pub).* 31 December 1997.

DA Pam 710-2-2. *Supply Support Activity Supply System: Manual Procedures.* 30 September 1998.

DA Pam 750-1. *Commanders' Maintenance Handbook.* 2 February 2007.

DA Pam 750-8. *The Army Maintenance Management System (TAMMS) Users Manual.* 22 August 2005.

**FIELD MANUALS**

FM 1-02 (FM 101-5-1). *Operational Terms and Graphics.* 21 September 2004.

FM 1-05. *Religious Support.* 18 April 2003.

FM 1-06 (FM 14-100). *Financial Management Operations.* 21 September 2006.

FM 3-11.3 (FM 3-3, FM 3-3-1, FM 3-22). *Multiservice Tactics, Techniques, and Procedures for Chemical, Biological, Radiological, and Nuclear Contamination Avoidance.* 2 February 2006.

FM 3-11.4. *Multiservice Tactics, Techniques, and Procedures for Nuclear, Biological, and Chemical (NBC) Protection.* 2 June 2003.

FM 3-11.5 (FM 3-5). *Multiservice Tactics, Techniques, and Procedures for Chemical, Biological, Radiological, and Nuclear Decontamination.* 4 April 2006.

FM 3-19.4. *Military Police Leaders' Handbook.* 4 March 2002.

FM 3-19.30 (FM 19-30). *Physical Security.* 8 January 2001.

FM 3-21.8 (FM 7-8). *Infantry Rifle Platoon and Squad.* 28 March 2007.

FM 3-21.71. *Mechanized Infantry Platoon and Squad (Bradley).* 20 August 2002.

FM 3-21.75 (FM 21-75). *The Warrior Ethos and Soldier Combat Skills.* 28 January 2008.

FM 3-25.26 (FM 21-26). *Map Reading and Land Navigation.* 18 January 2005.

FM 3-34.170 (FM 5-170). *Engineer Reconnaissance.* 25 March 2008.

FM 3-90.1 (FM 71-1). *Tank and Mechanized Infantry Company Team.* 9 December 2002.

FM 3-90.5 (FM 3-90.2). *The Tank and Mechanized Infantry Battalion Task Force.* 7 April 2008.

FM 3-100.12. *Risk Management for Multiservices Tactics, Techniques, and Procedures.* 15 February 2001.

FM 6-22. *Army Leadership.* 12 October 2006.

FM 7-22.7. *The Army Noncommissioned Officer Guide.* 12 December 2002.

FM 4-01.011 (FM 55-65 and FM 55-9). *Unit Movement Operations.* 31 October 2002.

FM 4-01.30. *Movement Control.* 1 September 2003.

FM 4-25.11 (FM 21-11). *First Aid.* 23 December 2002.

FM 4-30.3 (FM 9-43-1). *Maintenance Operations and Procedures.* 28 July 2004.

FM 5-0 (FM 101-5). *Army Planning and Orders Production.* 20 January 2005.

FM 5-19. *Composite Risk Management.* 21 Auggust 2006.

FM 5-34. *Engineer Field Data.* 19 July 2005.

FM 6-30. *Tactics, Techniques, and Procedures for Observed Fire.* 16 July 1991.

FM 7-7. *The Mechanized Infantry Platoon and Squad (APC).* 15 March 1985.

FM 7-22.7. *The Army Noncommissioned Officer Guide.* 23 December 2002.

FM 17-95. *Cavalry Operations.* 24 December 1996.

FM 21-60. *Visual Signals.* 30 September 1987.

FM 21-305. *Manual for the Wheeled Vehicle Driver.* 27 August 1993.

FM 27.2. *Your Conduct in Combat Under the Law of War.* 23 November 1984.

FM 27-10. *The Law of Land Warfare.* 18 July 1956.

FM 46-1. *Public Affairs Operations.* (*This Item is Included on EM 0205.*) 30 May 1997.

FM 55-1. *Transportation Operations.* 3 October 1995.

FM 55-15. *Transportation Reference Data.* 27 October 1997.

FM 55-30. *Army Motor Transport Units and Operations.* 27 June 1997.

**FORMS**

DA Form 6. *Duty Roster.*

DA Form 285. *U.S. Army Accident Report.*

DA Form 638. *Recommendation for Award.*

DA Form 1132-R. *Prisoner's Personal Property List - Personal Deposit Fund (LRA).*

DA Form 1156. *Casualty Feeder Card.*

DA Form 1971-10-R. *NBC-4 Radiation Doseate* Measurements/ *Chemical/Biological Areas of Contamination (LRA).*

DA Form 2166-8. *Noncommissioned Officer Evaluation Report.*

DA Form 2166-8-1. *Noncommissioned Officer Counseling and Support Form.*

DA Form 3964. *Classified Document Accountability Record.*

DA Form 4137. *Evidence/Property Custody Document.*

DA Form 7566. *Composite Risk Management Worksheet.*

DD Form 518. *Accident-Identification Card.*

DD Form 626. *Motor Vehicle Inspection (Transporting Hazardous Materials).*

DD Form 836. *Dangerous Goods Shipping Paper/Declaration and Emergency Response Information for Hazardous Materials Transported by Government Vehicles.*

DD Form 1265. *Request for Convoy Clearance.*

DD Form 1266. *Request for Special Hauling Permit.*

DD Form 2708. *Receipt for Inmate or Detained Person.*

DD Form 2745. *Enemy Prisoner of War [EPW] Capture Tag.*

SF 91. *Motor Vehicle Accident Report.*

SF 700. *Security Container Information.*

SF 702. *Security Container Check Sheet.*

SF 703. *Top Secret Cover Sheet.*

SF 704. *Secret Cover Sheet.*

SF 705. *Confidential Cover Sheet.*

**SOLDIER TRAINING PUBLICATIONS**

STP 21-1-SMCT. *Soldier's Manual of Common Tasks Warrior Skill Level 1. December 2007.*

**TECHNICAL MANUAL**

TM 3-4240-279-10. *Operator's Manual for Mask, Chemical-Biological: Field, ABC-M17 (NSN 4240-00-542-4450) Small; (4240-00-542-4451) Medium; (4240-00-542-4452) Large; M17A1 (4240-00-926-4199) Small; (4240-00-926-4201) Medium; (4240-00-926-4200) Large; M17A2 (4240-01-143-2017); X-Small (4240-01-143-2018); Small (4240-01-143-2019); Medium (4240-01-143-2020); Large.* 5 October 1987.

TM 3-6665-307-10. *Operator's Manual for Chemical Agent Detector Kit, M256 (NSN 6665-01-016-8399) and M256A1 (6665-01-133-4964).* 1 September 1985.

TM 3-9905-001-10. *Operator's Manual for Sign Kit, Contamination (NSN 9905-01-346-4716).* 23 August 1982.

TM 9-243. *Use and Care of Hand Tools and Measuring Tools.* 12 December 1983.

TM 10-8415-209-10. *Operator's Manual for Individual Chemical Protective Clothing.* 31 March 1993.

TM 43-0003-30. *Demilitarization Procedures for FSC 5180 Crimping Outfit and Maintenance Kit, FSC 6665 Hazard Detecting Instruments and Apparatus and FSC 6910 - Training Sets.* 30 November 1992.

**TRAINING CIRCULARS**

TC 1-05. *Religious Support Handbook for the Unit Ministry Team.* 10 May 2005.

TC 3-41. *Protection Assessment Test System* (PATS). 14 January 1995.

TC 27-10-1. *Selected Problems in the Law of War.* 26 June 1979.

**OTHER PUBLICATIONS**

FORSCOM Reg 55-2. *Unit Movement Data Reporting and System Administration.* 31 October 1997.

GTA 03-06-008. *CBRN Warning and Reporting System.* 1 November 2007.

JP 1-05. *Religious Support in Joint Operations.* 9 June 2004.

STANAG 2154. *Regulations for Military Motor Vehicle Movement by Road.* 19 June 1992.

STANAG 2155. *Road Movement Bid and Credit.* 26 August 1994.

STP 19-31B24-SM-TG. S*oldier's Manual and Trainer's Guide, MOS 31B, Military Police, Skill Levels 2/ 3/ 4.* 26 February 2007.

STP 19-31B1-SM. *Soldier's Manual, MOS 31B, Military Police, Skill Level 1.* 5 December 2007.

TB 55-46-1. *Standard Characteristics (Dimensions, Weight, and Cube) for Transportability of Military Vehicles and Other Outsize/Overweight Equipment (In Toe Line Item Number Sequence).* 1 January 2008.

## RELATED PUBLICATIONS

Related publications are sources of additional information. They are not required to understand this publication.

### ARMY REGULATIONS

AR 135-178. *Enlisted Administrative Separations.* 13 March 2007.

AR 190-8. *Enemy Prisoners of War, Retained Personnel, Civilian Internees and Other Detainees.* 1 October 1997.

AR 190-14. *Carrying of Firearms and Use of Force for Law Enforcement and Sec80-5urity Duties.* 12 March 1993.

AR 190-45. *Law Enforcement Reporting.* 30 March 2007.

AR 190-51. *Security of Unclassified Army Property (Sensitive and Nonsensitive).* 30 September 1993.

AR 350-1. *Army Training and Leader Development.* 3 August 2007.

AR 385-63. *Range Safety.* 19 May 2003.

AR 690-12. *Equal Employment Opportunity and Affirmative Action.* 4 March 1988.

AR 690-600. *Equal Employment Opportunity Discrimination Complaints.* 9 February 2004.

### DEPARTMENT OF THE ARMY PAMPHLETS

DA Pam 350-20. *Unit Equal Opportunity Training Guide.* 1 June 1994.

DA Pam 600-26. *Department of the Army Affirmative Action Plan.* 23 May 1990.

### FIELD MANUALS

FM 2-22.3 (FM 34-52). *Human Intelligence Collector Operations.* 6 September 2006.

FM 3-11.9. *Potential Military Chemical/Biological Agents and Compounds.* 10 January 2005.

FM 3-11.19. *Multiservice Tactics, Techniques, and Procedures for Nuclear, Biological, and Chemical Reconnaissance.* 30 July 2004.

FM 3-11.34. *Multiservice Tactics, Techniques, and Procedures for Installation CBRN.* 6 November 2007.

FM 3-11.86. *Multiservice Tactics, Techniques, and Procedures for Biological Surveillance.* 4 October 2004.

FM 3-19.1. *Military Police Operations.* 22 March 2001.

FM 3-19.13. *Law Enforcement Investigations.* 10 January 2005.

FM 3-19.15.*Civil Disturbance Operations*. 18 April 2005.

FM 3-19.40 (FM 19-40). *Internment/Resettlement Operations.* 4 September 2007.

FM 3-21.10 (FM 7-10). *The Infantry Rifle Company.* 27 July 2006.

FM 3-21.20 (FM 7-20). *The Infantry Battalion.* 13 December 2006.

FM 3-22.68 (FM 23-14). *Crew Served Weapons.* 21 July 2006.

FM 4-02.2. *Medical Evacuation.* 8 May 2007.

FM 4-02.4. *Medical Platoon Leaders' Handbook Tactics, Techniques, and Procedures.* 24 August 2001.

FM 4-02.17. *Preventive Medicine Services.* 28 August 2000.

FM 4-25.12. *Unit Field Sanitation Team.* 25 January 2002.

FM 5-102. *Countermobility.* 14 March 1985.

FM 6-0. *Mission Command: Command and Control of Army Forces.* 11 August 2003.

FM 6-22 (FM 22-100). *Army Leadership.* 12 October 2006.

FM 6-30. *Tactics, Techniques, and Procedures for Observed Fire.* 16 July 1991.

FM 7-7. *The Mechanized Infantry Platoon and Squad (APC).* 15 March 1985

FM 7-21.13. *The Soldier's Guide.* 2 February 2004.

FM 8-10-6. *Medical Evacuation in a Theater of Operations Tactics, Techniques, and Procedures.* 14 April 2000.

FM 8-55. *Planning for Health Service Support.* 9 September 1994.

FM 19-25. *Military Police Traffic Operations.* 30 Septemeber 1977.

FM 22-51. *Leaders' Manual for Combat Stress Control.* 29 September 1994.

FM 34-54. *Technical Intelligence.* 30 January 1998.

FM 46-1. *Public Affairs Operations.* 30 May 1997.

**FORMS**

DA Form 638. *Recommendation for Award.*

DA Form 2401. *Organization Control Record for Equipment.*

DA Form 2823. *Sworn Statement.*

DA Form 3881. *Rights Warning Procedure/Waiver Certificate.*

DA Form 3946. *Military Police Traffic Accident Report.*

DA Form 3975. *Military Police Report.*

DA Form 4002. *Evidence/Property Tag.*

DA Form 4137. *Evidence/Property Custody Document.*

DD Form 4. *Enlistment/Reenlistment Document Armed Forces of the United States.*

DD Form 1408. *Armed Forces Traffic Ticket (Book, Consisting of 25 Three-Part Sets).*

DD Form 1920. *Alcohol Incident Report.*

**SOLDIER TRAINING PUBLICATIONS**

STP 19-31B1-SM. *Soldier's Manual, MOS 31B, Military Police, Skill Level 1.* 5 December 2007.

**TRAINING CIRCULAR**

TC 26-6. *Commander's Equal Opportunity Handbook.* 23 June 2008.

**TECHNICAL MANUAL**

TM 3-4240-279-10. *Operator's Manual for Mask, Chemical-Biological: Field, ABC-M17.* 5 October 1987.

TM 3-4240-279-20&P. *Unit Maintenance Manual (Including Repair Parts and Special Tools List) for Mask, Chemical-Biological; Field, ABC-M17 (NSN 4240-00-542-4450), (Small), (4240-00-542-4451), (Medium) and (4240-00-542-4452), (Large); M17A1 (4240-00-926-4199), (Small), (4240-00-926-4201), (Medium) and (4240-00-926-4200) (Large) and M17A2 (4240-01-143-2017), (X-Small), (4240-01-143-2018), (SMALL), (4240-01-143-2019), (Medium) and (4240-01-143-2020).* 5 October 1987.

**OTHER PUBLICATIONS**

Center for Health Promotion & Preventive Medicine (CHPPM) Tobacco

*Note:* CHPPM publications are at http://www.hooah4health.com/4you/stoptobaccoshop/default.htm?f=targeting.htm

FORSCOM Reg 55-1. *Unit Movement Planning.* 1 June 2006.

***Note***. FORSCOM publications are at http://www.forscom.army.mil/pubs/

MCM. *Manual for Courts-Martial.* 1 July 2005

*Note:* This publication can be found at https://akocomm.us.army.mil/usapa/epubs/27_Series_Collection_1.html

USACHPPM Technical Guide (TG) 240, *Combat Stress Behaviors.*

USACHPPM TG 241, *Combat & Operational Stress Reactions.*

USACHPPM TG 242. *Battle Fatigue/Combat Stress Reaction Prevention: Leader Actions.*

## WARRIOR ETHOS

*The Warrior Ethos forms the foundation for the American Solider's spirit and total commitment to victory, in peace and war, always exemplifying the ethical behavior and Army Values. Soldiers put the mission first, refuse to accept defeat, never quit, and never leave behind a fellow American. Their absolute faith in themselves and their comrades makes the United States Army invariably persuasive in peace and invincible in war.*

30794517R00163

Made in the USA
Columbia, SC
30 October 2018